CHAOTIC PHENOMENA IN ASTROPHYSICS

ANNALS OF THE NEW YORK ACADEMY OF SCIENCES
Volume 497

CHAOTIC PHENOMENA IN ASTROPHYSICS

Edited by J. Robert Buchler and Heinrich Eichhorn

The New York Academy of Sciences
New York, New York
1987

Copyright © 1987 by The New York Academy of Sciences. All rights reserved. Under the provisions of the United States Copyright Act of 1976, individual readers of the Annals are permitted to make fair use of the material in them for teaching and research. Permission is granted to quote from the Annals provided that the customary acknowledgment is made of the source. Material in the Annals may be republished only by permission of The Academy. Address inquiries to the Executive Editor at The New York Academy of Sciences.

Copying fees: For each copy of an article made beyond the free copying permitted under Section 107 or 108 of the 1976 Copyright Act, a fee should be paid through the Copyright Clearance Center Inc., 21 Congress St., Salem, MA 01970. For articles of more than 3 pages the copying fee is $1.75.

Cover: The cover shows a histogram from a periodic 2,764,949-iteration orbit (see page 63).

Library of Congress Cataloging-in-Publication Data

Chaotic phenomena in astrophysics.
  (Annals of the New York Academy of Sciences, ISSN 0077-8923: v. 497)
  "The papers in this volume were presented at a workshop entitled Chaotic phenomena in astrophysics, which was held at the University of Florida at Gainesville, on October 9–10, 1986 ... constituted the proceedings of the Second Florida Workshop in Nonlinear Astronomy"—P.
  Bibliography: p.
  Includes index.
  1. Chaotic behavior in systems—Congresses. 2. Astrophysics—Congresses. I. Buchler, J. R. (J. Robert).
II. Eichhorn, Heinrich K. (Heinrich Karl), 1927– .
III. Florida Workshop in Nonlinear Astronomy (2nd: 1986: University of Florida). IV. Series.
Q11.N5   vol. 497   500 s   87-12324
[QB466.C45]   [523.01]
ISBN 0-89766-389-6
ISBN 0-89766-390-X (pbk.)

SP
Printed in the United States of America
ISBN 0-89766-389-6 (cloth)
ISBN 0-89766-390-X (paper)
ISSN 0077-8923

ANNALS OF THE NEW YORK ACADEMY OF SCIENCES

Volume 497
May 29, 1987

CHAOTIC PHENOMENA IN ASTROPHYSICS[a]

Editors
J. ROBERT BUCHLER and HEINRICH EICHHORN

CONTENTS

Introduction. *By* JEAN-ROBERT BUCHLER and HEINRICH EICHHORN	ix
Stochasticity in Galactic Models. *By* G. CONTOPOULOS	1
Galactic Models with Moderate Stochasticity. *By* MARTIN SCHWARZSCHILD	16
The Quadratic Zeeman Effect in Moderately Strong Magnetic Fields. *By* SHANNON L. COFFEY, ANDRÉ DEPRIT, BRUCE MILLER, and CAROL A. WILLIAMS	22
Chaotic Behavior in Variable Stars. *By* J. ROBERT BUCHLER	37
Chaos and the Solar Cycle. *By* E. A. SPIEGEL and ALAN WOLF	55
Strange Accumulators. *By* L. A. SMITH and E. A. SPIEGEL	61
On the Origin of Large-Scale Cosmological Structure. *By* J. N. FRY	66
Stability of an Area-Preserving Mapping. *By* JAMES H. BARTLETT	78
Resonances Fill Stochastic Phase-Space. *By* J. D. MEISS	83
Hamiltonian and Dissipative Chaos. *By* GEORGE SCHMIDT	97
The Breakdown of KAM Trajectories. *By* D. BENSIMON and L. P. KADANOFF	110
Fractal Basin Boundaries with Unique Dimension. *By* CELSO GREBOGI, EDWARD OTT, JAMES A. YORKE, and HELENA E. NUSSE	117
Applications of the Semiclassical Spectral Method to Nuclear, Atomic, Molecular, and Polymeric Dynamics. *By* M. L. KOSZYKOWSKI, G. A. PFEFFER, and D. W. NOID	127
Kelvin-Helmholtz Instabilities in the Interstellar Medium. *By* JAMES H. HUNTER, JR. and RODNEY W. WHITAKER	144
Index of Contributors	155

---

[a]The papers in this volume were presented at a workshop entitled Chaotic Phenomena in Astrophysics, which was held at the University of Florida at Gainesville, on October 9–10, 1986. The presentation of these papers constituted the Proceedings of the Second Florida Workshop in Nonlinear Astronomy.

The New York Academy of Sciences believes it has a responsibility to provide an open forum for discussion of scientific questions. The positions taken by the participants in the reported conferences are their own and not necessarily those of the Academy. The Academy has no intent to influence legislation by providing such forums.

Ante mare et terras et, quod tegit omnia, caelum
Unus erat toto naturae vultus in orbe,
Quem dixere Chaos; . . .

Metamorphoseon I, 5–7
Publius Ovidius Naso

Introduction

JEAN-ROBERT BUCHLER[a] AND HEINRICH EICHHORN[b]

[a]*Department of Physics*
[b]*Department of Astronomy*
University of Florida
Gainesville, Florida 32611

The Second Florida Workshop on Dynamical Astronomy was held October 9–10, 1986 at the University of Florida at Gainesville. Organized under the leadership of George Contopoulos, it was devoted to chaotic phenomena in astrophysics and brought together a panel of invited speakers of widely varied backgrounds who, in the course of working on problems in their fields (ranging from the behavior of molecules to the structure of galaxies), had encountered the need to study the general behavior of dynamical systems.

It may be appropriate to remind ourselves that the researches reported on, in the present collection, are in a field that has the appearance (on the time scale of human endeavors) as if it were entirely new. Let an example illustrate this: the Viennese astronomer, Oswald Thomas, wrote (in 1947) a popular brochure entitled "Fahrt zum Mond" ("Trip to the Moon") and therein had a fictitious questioner ask (less than a quarter of a century before man landed on the moon and went back to Earth): "Do you think a trip to the moon is at all possible?" Thomas counters by asking: "Is it at all possible that an apple tree, vigorously shaken, will shed fried pears?" He then proceeds to point out that fried pears may very well fall from a shaken apple tree if they have previously all been tied to it by fragile threads. He ends his argument by pointing out how unlikely it is that one will find too many apple trees with fried pears tied to them.

By and large, Thomas' attitude reflected the opinion of the astronomical community at the time and, in particular, that of the community of celestial mechanicians. After Bruns's proofs of the nonexistence of algebraic integrals beyond those then already known and Poincaré's of the nonexistence of further uniform integrals of the three-body problem in the second half and around the turn of the nineteenth century, respectively, the celestial mechanicians had to abandon the hope of finding a closed solution of the three- (and, *a fortiori*, the multi-) body problem. (It is unfortunate that many of the widely used texts on mechanics mislead the readers into thinking that integrable Hamiltonians are the rule, while, in reality, they are a subset of measure zero of the set of all Hamiltonians.) The research in the field then concentrated even more on the construction of convergent (so one always hoped) analytical approximations. The disadvantage of these is that they describe reality accurately only for a time span that is short in comparison with the age of the planetary system. These analytical developments are therefore of no help in the quest for obtaining information about the long-term behavior of the bodies of the solar system, let alone for shedding any light on cosmogonic questions. It was clear that the integration by analytic approximations of the differential equations of planetary motion would not throw much light on the general properties of the solutions.

The fact that celestial mechanicians had been able to compute trajectories in spite of the appearance of unavoidable divergent denominators in the perturbation expansions eventually led Kolmogorov, Arnold, and Moser (around 1960) to their famous KAM theorem that, while throwing a monkey wrench into statistical mechanics where many physicists had come to take ergodicity for granted, showed that a large number of periodic orbits can survive in spite of nonintegrability of their Hamiltonians.

Jumping back in time, we recall that significant progress, in view of this discouraging situation, came principally from two scholars: Henri Poincaré, himself, and Elis Strømgren. Around the turn of the last century, Poincaré in his profound "Les Méthodes Nouvelles de la Mécanique Céleste" started investigating periodic orbits and their stability, which involved the consideration of, as we would say today, the topology of the phase-space of the orbits. He introduced the "surfaces of section", subspaces of phase-space, in which the plot of the points corresponding to various individual orbits can give profound insights into the character of the orbits, provided that one knows the trajectory itself. The utilization of Poincaré's insight therefore had to lie dormant for half a century until numerical integration on fast electronic computers rendered the finding of the actual trajectory a performable routine.

In the meantime—mostly, in the 1920s—Elis Strømgren in Copenhagen had succeeded to a large extent in exploring numerically the circular restricted three-body problem. He did this basically by farming out the numerical integration (by means of desk calculators) of the equations of motion to his graduate students. Almost all we knew about the restricted three-body problem had been found out by the researchers of the Copenhagen School. It was on the basis of these that Thomas had made his pronouncements. In 1947, the electronic computer, now an item taken for granted by every undergraduate, existed only in the brains of some daring designers. The timely calculations of a midcourse correction of a space vehicle could not have been accomplished by desk calculators, and even the computation of the actual trajectory of a space vehicle to the moon would have been a major undertaking on a desk calculator. From what Thomas knew in 1947, a spacecraft might safely have been launched from the Earth, but to land it safely on the moon, without assistance from an electronic computer, was in fact impossible. We would like to predict that the historians of the future will recognize the period after 1950 as a new epoch in the history of humanity, and if named properly, it will aptly be named the Computer Age rather than the (restrictive) Atomic Age.

One of us (HKE) remembers attending a talk given by G. M. Clemence in 1958 in which he discussed computers. This was after computers had been successfully mass-produced and an IBM 650 (remember those?) had come within reach of every serious researcher. His opinion was then that one hundred IBM 650s would more than satisfy the needs of the astronomical community in the United States for many years.

Matters, though, changed soon. The computations of the location and the ephemerides of the bodies in the planetary system became routine problems that would now be solved by the simultaneous numerical integration of the relevant differential equations. The numerical integration of the equations of motion in the general few-body problem for varied sets of initial conditions was a significant piece of work in the early computer age. Now, in general, the community of mainstream celestial mechanicians and dynamicists no longer regards such work worth publishing. Instead, the results of numerical experiments now communicate surfaces of sections, stability diagrams, and existence diagrams, in which one point represents an orbit calculated with a set of initial conditions or even a family of such initial conditions. No longer does the main quest concern individual orbits, but rather it is concerned with the topological properties of the phase-space associated with a particular Hamiltonian, the stability of whole families of orbits, and such. Numerical integration on large computers has rendered the calculation of orbits for a particular set of initial conditions a routine matter, and surprises abound—for example, the trajectories of the vehicles that traveled from the Earth to the moon are not in any of the families explored by the Copenhagen school.

The impetus for the whole new numerical study was given in 1964 by the famous work of Hénon and Heiles on a nonintegrable Hamiltonian with a cubic potential that exhibits large regions of stochasticity. Concurrently with Hénon and Heiles, Lorenz in 1963 (as well as Moore and Spiegel in 1966) initiated the study of dissipative systems that exhibit stochastic or chaotic behavior with his model equations for convection in the atmosphere. (Not surprisingly, Poincaré already pioneered this area as well.) In many problems of planetary, stellar, and galactic dynamics, dissipation seems to play a minor role, and Hamiltonian studies provide an excellent description. However, in those areas where fluid dynamics *largo sensu* (e.g., stellar pulsations, accretion problems, dynamo theory) is involved, dissipation generally has important consequences on the time scale of interest. It is therefore not astonishing that the importance of the study of both Hamiltonian and dissipative dynamics has been realized in astrophysics.

The variety of the backgrounds of the participants in this conference and the authors of the papers in these proceedings is a fine example for "interdisciplinary" science. It emerged because the scientists themselves perceived its necessity. For this reason, it will prosper.

The thanks of the editors and the participants in the conference are due to the Departments of Astronomy and of Physics at the University of Florida, whose moral and financial support has made this gathering not only possible, but highly successful.

Stochasticity in Galactic Models

G. CONTOPOULOS

Department of Astronomy
University of Florida
Gainesville, Florida 32611
and
Department of Astronomy
University of Athens
Athens, Greece

INTRODUCTION: THE ONSET OF LARGE DEGREE STOCHASTICITY

The present Workshop on Chaotic Phenomena in Astrophysics is devoted to the role of "deterministic chaos" in various astrophysical problems. The term, "deterministic chaos", means that chaos is not introduced a priori, but results from solutions of differential equations, or successive iterations of deterministic mappings.

In the general case of time-independent Hamiltonian systems of two degrees of freedom, the change of a parameter (e.g., the energy) changes the motions from ordered (quasi-periodic) to stochastic (chaotic) (FIGURES 1a and 1b).

The transition from a nearly integrable system to a chaotic system is rather well understood today. A nearly integrable system has a large set of ordered motions that fill toroidal surfaces in phase-space (KAM tori). The system contains an infinite number of stable and unstable periodic orbits, and it is known that there is a small set of stochastic (chaotic) motions near every unstable orbit. These chaotic motions, though, are well separated by KAM tori (FIGURE 2a). The intersections of the KAM tori by a Poincaré surface of section are closed invariant curves. As the energy increases, the chaotic regions increase. However, only after the destruction of the "last KAM curve" do they communicate and introduce a large degree of stochasticity. The last KAM curve has a "noble" rotation number that can be written in the form of a continuous fraction:

$$\text{rot} = \cfrac{1}{a_1 + \cfrac{1}{a_2 + \cfrac{1}{a_3 + \cdots}}} \qquad (1)$$

with $a_i = 1$ for all i larger than a certain order N. The evolution of the last KAM curve has been described by Greene[1] and by Shenker and Kadanoff[2] (FIGURE 2). As the energy (or an equivalent nonlinearity parameter) increases, there appear families of periodic orbits (on both sides of the noble invariant curve) that correspond to the successive truncations of the noble rotation number. At the same time, the noble invariant curve develops corrugations to make room for the stable periodic orbits and their neighborhoods (FIGURE 2b). As we approach a critical value, h_{crit}, the number of the periodic orbits tends to infinity and all periodic orbits close to the noble invariant

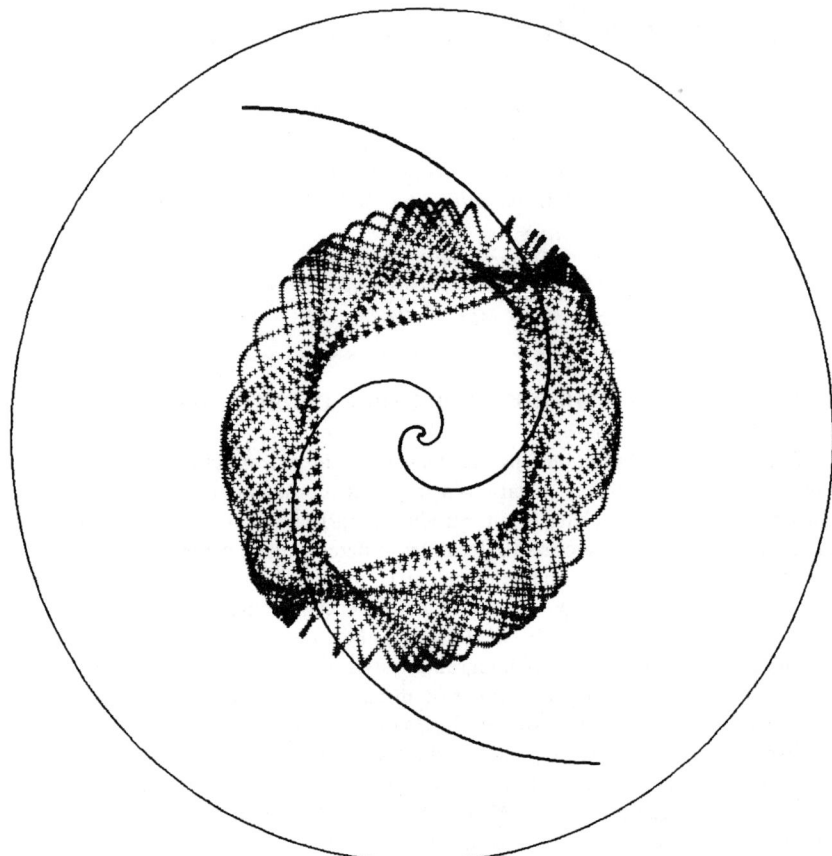

FIGURE 1a. An ordered (quasi-periodic) orbit in a spiral galaxy.

curve become unstable. At the critical value itself, the noble invariant curve has corrugations on all scales; that is, it is a fractal. This is the last KAM curve. One can describe its structure by a renormalization theory.[2,3]

As we go to a value of $h > h_{crit}$, the noble invariant curve develops an infinite number of holes that form a Cantor set. Thus, it is called a "cantorus."[4-7] The chaotic regions on both sides of the cantorus can now communicate and produce a large connected chaotic region (FIGURE 2c). The diffusion through the holes of the cantorus for h slightly larger than h_{crit} has been estimated theoretically by Bensimon and Kadanoff[8] and by MacKay, Meiss, and Percival.[9]

A similar theory can be developed in finding the boundaries of various "islands", that is, sets of invariant curves around a periodic orbit. The outermost invariant curve is a fractal, surrounded by cantori and islands corresponding to higher order resonances. The outward diffusion of orbits starting close to this fractal is initially very slow, but later on it is faster. We observed this phenomenon in a galactic problem several years ago[10] (FIGURE 3), but it was later emphasized by a number of authors[11-13] who spoke about "vague tori" or about the "stickiness" property of the invariant curves.

A practical way of finding the last invariant curve is based on a conjecture by Greene[1] that the periodic orbits corresponding to the successive approximations of a noble number become unstable as we approach the critical energy (or critical nonlinearity). Schmidt and Bialek[14] developed a numerical method for finding the limits of the stochastic regions for any value of the nonlinearity parameter by determining the transition to instability of several periodic orbits close to this limit.

Most of the numerical studies of the transition problem have been made with mappings. In the case of a galactic problem, the calculation of successive Poincaré consequents on a surface of section is much more time-consuming. On the other hand, the number of periods available during a Hubble time (the "age of the universe") is relatively small, specifically, of the order of 100.

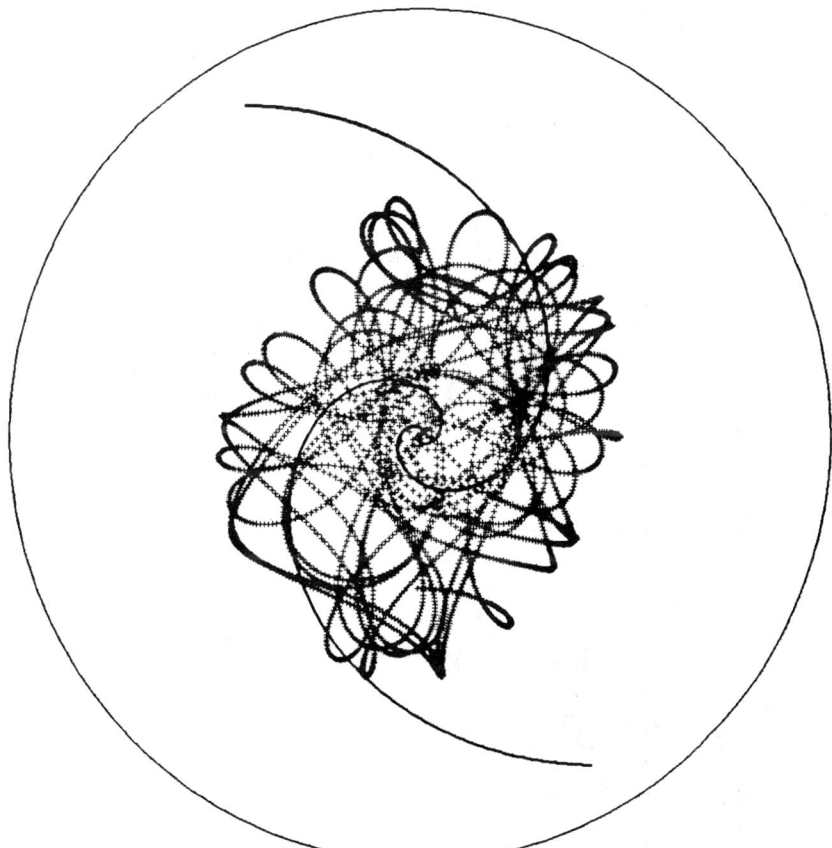

FIGURE 1b. A stochastic orbit in a spiral galaxy.

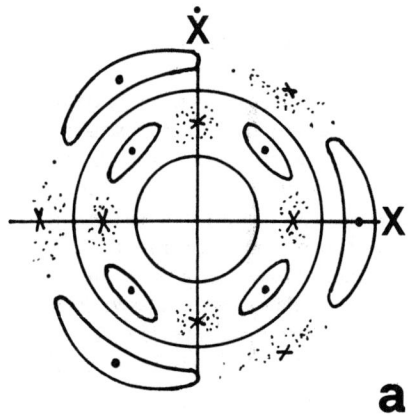

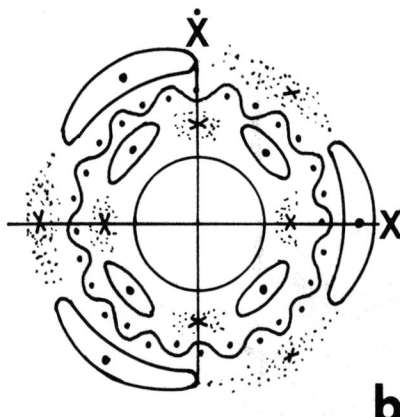

FIGURE 2. The destruction of a noble invariant curve. (a) For small h, the noble curve is surrounded by a set of KAM curves. Outside and inside these curves are sets of islands surrounding invariant points of orders 3 and 4 (corresponding to triple and quadruple stable periodic orbits, respectively). Between the islands are unstable invariant orbits surrounded by chaotic orbits; however, the chaotic regions are small and well separated. (b) As h approaches h_{crit}, the noble invariant curve develops corrugations and new sets of invariant points appear on both its sides. (c) For $h > h_{crit}$, the noble invariant curve develops an infinite number of holes (cantorus) and the stochastic regions inside and outside it communicate.

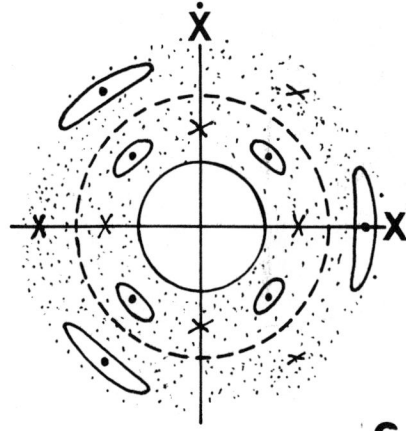

If the diffusion through a cantorus is effective over a time scale much larger than the age of the universe, we can ignore it and consider the cantorus as a real barrier. For example, a cantorus may not allow the escape of a star from a galaxy within a Hubble time even if its energy is larger than the escape energy, h_{esc}.

In the other extreme case of a fast diffusion (if the energy h is appreciably larger than h_{crit}), we can ignore the existence of the cantorus altogether.

Thus, in galactic dynamics, we need the following information:

(a) The approximate positions of the main cantori and the corresponding critical energies, h_{crit}.
(b) To find for what energies, above h_{crit}, the diffusion rates through cantori become appreciable.
(c) To find the effects of cantori on the escape rate. More generally, we want to find when a star, with energy larger than h_{esc}, is expected to escape from a galaxy.
(d) Finally, we want to know what changes are introduced if the galaxy is three-dimensional and not flat.

The following sections discuss these problems in some simple galactic models.

THE TRANSITION TO STOCHASTICITY IN 2-D GALACTIC MODELS

We consider a simple model representing the central region of a deformed nonrotating galaxy, that is, given by the Hamiltonian

$$H \equiv \tfrac{1}{2}(\dot{x}^2 + \dot{y}^2 + x^2 + y^2) + xy^2 = h. \qquad (2)$$

For small energies h, the main periodic orbits are:[15]

(A) an unstable orbit intersecting perpendicularly the x-axis,
(B) a stable, nearly circular orbit around the origin,
(C) two stable rectilinear orbits through the origin, and
(D) the axis $y = 0$ (stable).

As h increases, the family C becomes unstable at $h = h_1$, then stable again at $h = h_2$, and so on. As h approaches the escape energy, h_{esc}, it has an infinite number of transitions to stability and instability.[16,17]

At particular values of h, various resonant families of periodic orbits or invariant curves with irrational rotation numbers bifurcate from C (FIGURE 4). As h increases beyond certain critical values, h_{crit}, these invariant curves become cantori. In FIGURE 4, we show by straight lines the intervals of existence of noble invariant curves [3,1,1,...], [2,1,1,...], and [1,1,1,...]. We see that the invariant curve [3,1,1,...] starts earlier (at smaller h) and ends a little later than the invariant curve [2,1,1,...]. On the other hand, the "golden mean" invariant curve [1,1,1,...] first appears at a relatively large h and its extent in h is small.

For h smaller than h_1, the orbit C is stable (FIGURE 5, for $h = 0.111$). Around the point A, representing the unstable orbit A, there is an appreciable stochasticity. Some stochasticity appears also between the islands surrounding the point C (e.g., the 3 and 7

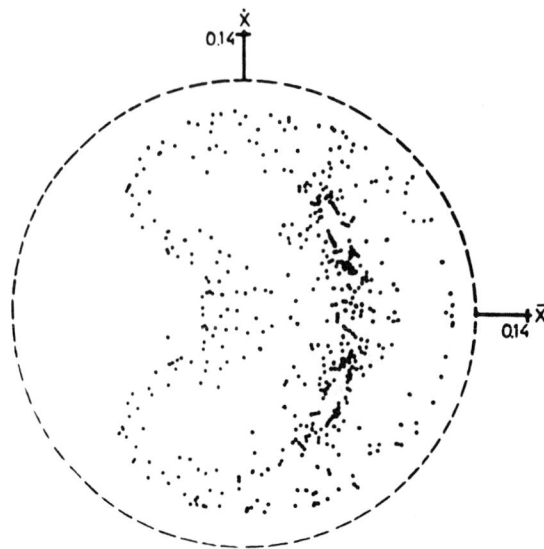

FIGURE 3. The consequents of a stochastic orbit starting close to one of the "last invariant curves" surrounding two symmetric islands (after Contopoulos[10]). Many consequents stay close to the last invariant curves for a long time, but later they diffuse outwards.

islands of FIGURE 5), but these stochastic regions are well separated from each other and from the stochastic region around A by closed invariant curves.

As h increases ($h = 0.115$, FIGURE 6), the boundaries between the various islands (e.g., the 3 and outer 7 islands of FIGURE 6) disappear, but there is still a closed invariant curve separating the inner stochastic region from the stochastic region around A. This case is very close to the critical value h_{crit} for the noble number $[3,1,1,\ldots]$. The corresponding last KAM curve is approximately the outer boundary of the stochastic region.

For even larger h, this invariant curve becomes a cantorus. If h is sufficiently larger than h_{crit} ($h = 0.125$, FIGURE 7), the holes of the cantorus are large and there is practically no retardation of the diffusion through it. In FIGURE 7, several other cantori with small holes appear closer to the point C and around some islands. All the points of FIGURE 7 are successive Poincaré consequents of only one initial point, which is in the dark region beyond the central hole. The regions of different density of points are separated by cantori with small holes, which provide partial barriers for long times, but not for all times. We see that many important cantori coexist in the same system.

In order to find a particular last KAM curve, as we do not know its exact position, we have to calculate many orbits for a long time and for several values of h so that we can find the value of h where we first see an indication of diffusion through a cantorus. A much more economical way of finding h_{crit} is to find when periodic orbits close to the last KAM curve become unstable.[18] In the case of the noble rotation number $[3,1,1,\ldots]$, the successive truncations of the continuous fraction in equation 1 give the numbers, $1/3, 1/4, 2/7, 3/11, 5/18, 8/29, \ldots$, alternately above and below the noble number. The corresponding periodic orbits bifurcate from C at relatively small values of h and are stable up to certain values of h (marked by the dots in FIGURE 8). If we join, by a straight line, two successive points above the noble number, or two points below it, we find, by extending this line until it intersects the straight line rot = [3,1,1, . . .], the approximate value of h_{crit} (a little above $h = 0.116$). We notice that the transition values of h become smaller as the order of the truncation increases.

We have verified this behavior in several cases. Thus, it is sufficient to calculate only two families of periodic orbits on one side (above or below) of the noble number and to join the points (h, rot) where these families become unstable in order to find approximately, by extrapolation, the critical value h_{crit}.

At the escape energy, $h = h_{esc} = 0.125$, all stochastic regions around C are connected, but there are still cantori providing partial barriers (FIGURE 7).

If we increase the energy beyond h_{esc}, the orbit C extends to infinity and there is a region around C in which the orbits escape quite fast (from a few periods up to a few

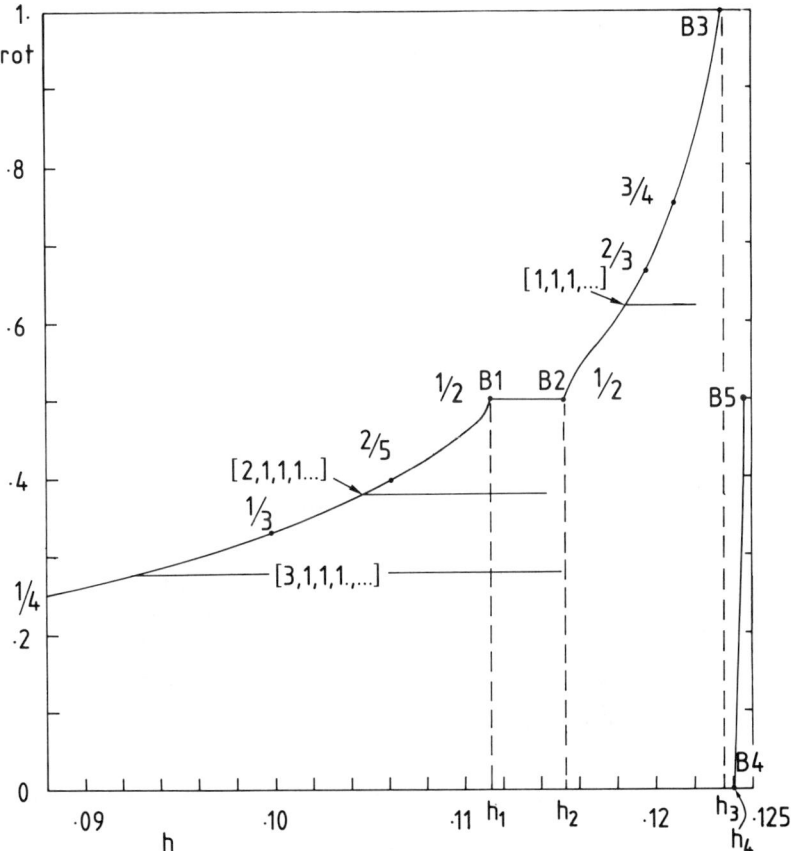

FIGURE 4. The rotation number (rot) of the periodic orbit C as a function of the energy (h). The rational numbers indicate the periodic orbits bifurcating from C. The orbit C is unstable in the intervals (h_1, h_2), (h_3, h_4), etc. At the points $B1$, $B2$, and $B5$, double periodic orbits bifurcate, and at the points $B3$ and $B4$, simple periodic orbits bifurcate. The horizontal lines with "noble" rotation numbers mark the intervals of existence of the corresponding invariant curves.

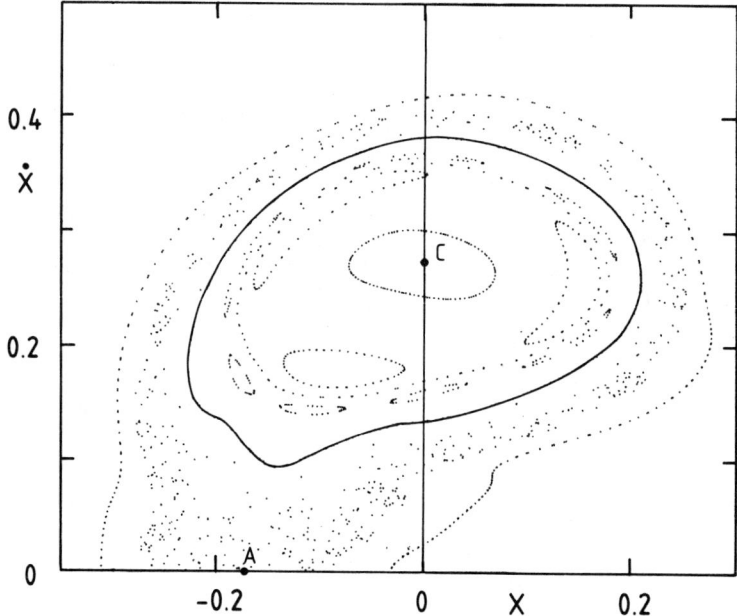

FIGURE 5. For relatively small h ($h = 0.111$), the various stochastic regions on a surface of section are well separated from each other and from the large stochastic region around the unstable invariant point A. The approximate position of the invariant curve [3,1,1,...] is marked.

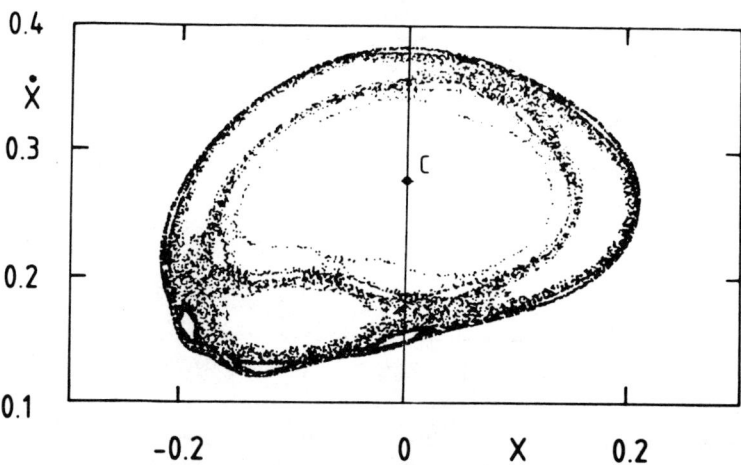

FIGURE 6. Ten thousand consequents of a stochastic orbit for $h = 0.115$. The stochastic region surrounding C is separated from the stochastic region around A by a noble invariant curve [3,1,1,1,...]. In this case, h is a little smaller than h_{crit}.

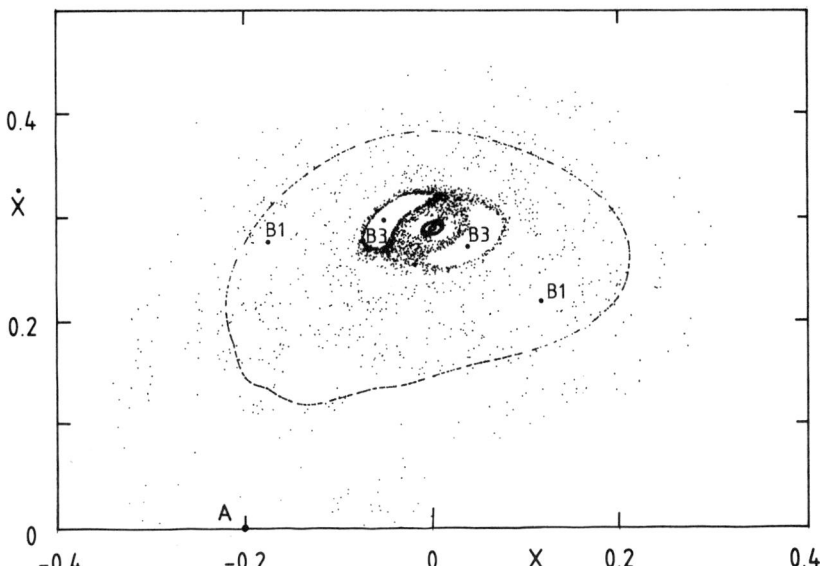

FIGURE 7. For h much larger than h_{crit} ($h = 0.125$), the cantorus [3,1,1,...] does not provide an effective barrier. The approximate position of the cantorus is given, but the holes are larger than shown. The change of density of points (regions of different darkness) indicates the existence of further cantori. $B1$ and $B3$ are stable periodic orbits of double and equal period.

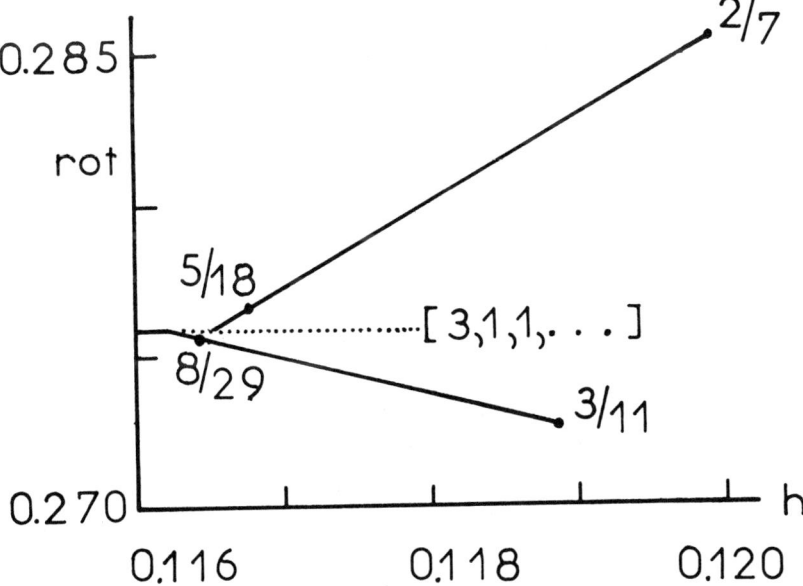

FIGURE 8. How to find the last KAM torus (Contopoulos et al.[18]): We mark the points (h = energy, rot = rotation number) where the periodic orbits corresponding to successive truncations of the noble number [3,1,1,1,...] become unstable. Extrapolating the line (2/7, 5/8) until it reaches the line [3,1,1,...], we find approximately the critical value of $h = h_{crit}$. The same is done by extrapolating the line (3/11, 8/29).

hundred periods). Orbits that start further out diffuse initially outwards (FIGURE 9). They stay for a somewhat long time in the darker region surrounding C and then they pass into the large stochastic region above and below A, moving several times up and down the axis $\dot{x} = 0$. The main cantori in this case consist of the one surrounding the darker region around C and a symmetric cantorus around C'. The orbit was calculated for 10,000 periods and did not pass again through these cantori. Of course, it is expected that the orbit will not only pass again through one of these cantori, but that it will eventually escape to infinity through the opening near C, or near C'. However, this may happen after a much longer time than 10^4 periods.

The question of time scales is very important in astronomy. In fact, the time intervals of interest should not be longer than the age of the universe, which is of the

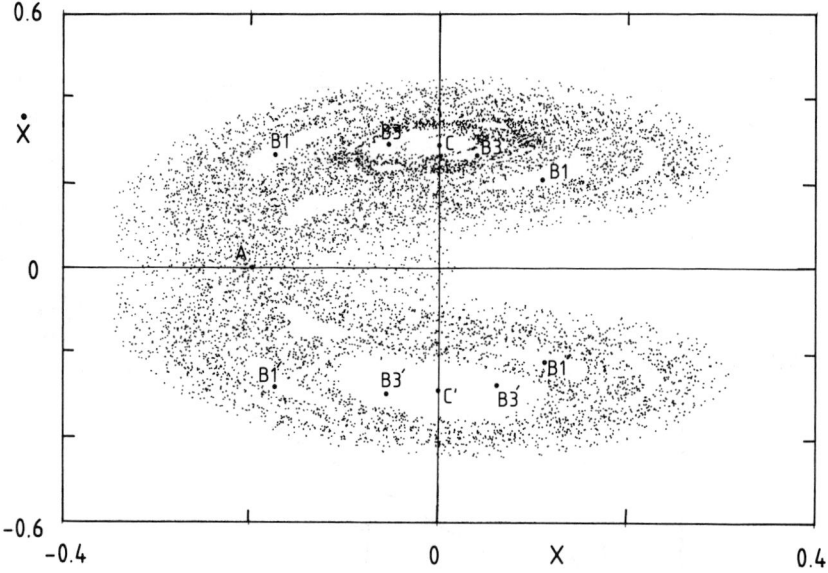

FIGURE 9. Ten thousand consequents of an orbit starting near (but not very close) to C in the case of $h = 0.12505$ (above the escape energy of $h_{esc} = 0.125$).

order of 10^{10} years. In galactic dynamics, the period of a star moving around the galaxy is of the order of 10^8 years; therefore, a Hubble time contains only about 100 periods. This implies that some cantori act as effective barriers, that is, like invariant curves. Although some stellar orbits can diffuse through them, the number of diffusing stars is so small that they can be safely ignored. In general, only a few cantori are "effective" in galactic models. One can identify them by calculating a number of orbits with various initial conditions over a Hubble time. Then, we see that certain orbits fill stochastically a region between two cantori or between the last invariant curve surrounding an island and a cantorus. This phenomenon has been discussed in some cases by Petrou.[19]

In the solar system, on the other hand, the periods of planets are of the order of

1–10 years; thus, a Hubble time contains 10^9–10^{10} periods, and diffusion through cantori may play a much more important role.

In the case of a plasma, a typical period is of the order of 10^{-6} seconds. Therefore, during a plasma confinement time of one minute, we have about 10^8 periods. Here again, the diffusion through cantori may be important.

Thus, in the cases of the solar system and of a plasma, much more numerical work is needed in order to find out the importance of cantori in the diffusion process of stochastic orbits.

On the other hand, the fact that the time scales are so short in terms of periods in galactic dynamics indicates that we can find in the motions of stars (or galaxies) the imprints of the initial conditions during the formation of galaxies (or clusters of galaxies).

ESCAPES

The escape of stars from a stellar system is an important subject for astronomy. It is usually assumed that stars with energy larger than the "escape energy" escape from the system. However, this assumption is not correct as we have seen above. Although in such a case the energy surface extends to infinity and some stars do escape, there may in fact be other restrictions that do not allow a particular set of stars to escape. For example, a stable periodic orbit and orbits in its neighborhood do not escape.

We have found[20,21] that an escape region in the plane (h, x) (where h is the energy and x is the distance from the center of an orbit starting with $y = \dot{x} = 0$) is surrounded by an infinite number of families of resonant periodic orbits. This can be understood easily in the case of a rotating spiral galaxy. In this case, the "central" family x_1 of periodic orbits (the circular orbits of the axisymmetric case) is broken by an infinity of gaps (FIGURE 10a) (at every even resonance

$$\frac{\kappa}{\Omega - \Omega_s} = \frac{2n}{1},$$

where κ is the epicyclic frequency, Ω is the angular velocity around the center, and Ω_s is the angular velocity of the system). As $n \to \infty$, the curves of zero velocity (equipotentials in the rotating system) open at the Lagrangian points, L_1 and L_2, and some stars escape. The corresponding $h = h_{esc}$ is the escape energy in the rotating frame. At this value of h, there starts an "escape region", that is, a set of initial conditions x that lead to escape. As the perturbation (e.g., the amplitude of a spiral) increases, the gaps increase and the resonant orbits have only small stable parts near their minimum h (FIGURE 10b).

Normally, the escape regions extend to infinity in the h direction. If, however, the given stellar system is inside a larger system, there may be no escapes until h becomes much larger than h_{esc}. Barbanis[21,22] studied the model Hamiltonian

$$H = \tfrac{1}{2}(\dot{x}^2 + \dot{y}^2 + x^2 + y^2) - xy^2 + \alpha y^4 \tag{3}$$

for various values of α. He found that for $\alpha \gtrsim 0.28$, the main escape region does not extend to infinity, while for $\alpha \simeq 0.29$, this escape region shrinks to one point. If $\alpha \geq 0.5$, there is no escape region at all and the escape energy is infinite.

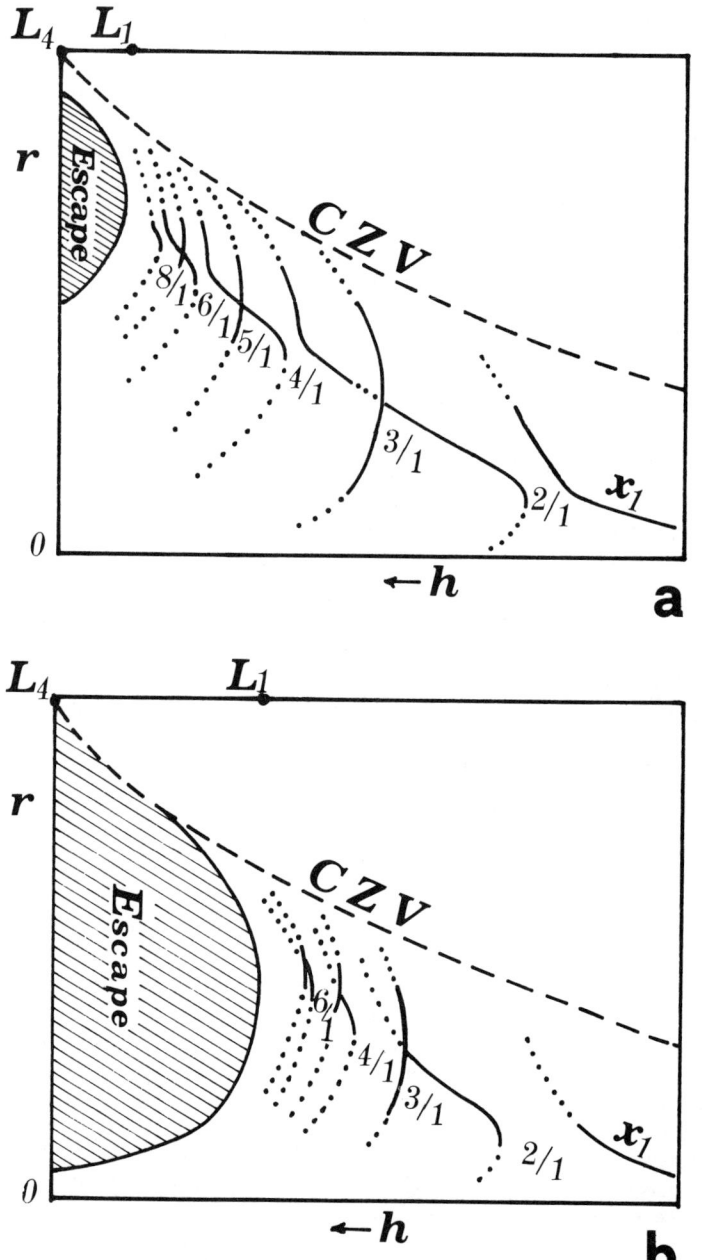

FIGURE 10. The escape regions in two rotating spiral galaxies (hatched): (a) weak spiral; (b) strong spiral. Included in the diagrams are the characteristics of stable (——) and unstable (· · ·) periodic orbits (distance from center r versus the energy h in the rotating frame). The dashed line (– – –) is the curve of zero velocity, and L_4 (stable) and L_1 (unstable) are Lagrangian points.

When the escape region is finite, it is surrounded on all sides by an infinite number of families of periodic orbits. Some of them form closed bubbles around the escape region, while the outermost family forms a finite number of loops spiraling inwards and then again outwards, until it joins some other family outside the region of the spirals. As α goes beyond $\alpha \simeq 0.29$ and the main escape region disappears, the number of bubbles becomes finite and the number of spiraling loops decreases. One after the other, the central loops disappear by shrinking to one point. For $\alpha = 0.5$, we have 8 double bubbles at the center and 16 spiraling loops, which are connected by 16 more bubbles.

These examples show us the structure of the neighborhood of the escape regions and they help us to find the proportion of escaping stars.

We may add that not only regions of islands around stable periodic orbits do not escape, but also regions restricted by cantori may not escape during a Hubble time. Thus, the escape rate is smaller than if we assume that all stars with $h > h_{esc}$ escape.

THREE-DIMENSIONAL GALACTIC MODELS

The inclusion of a third dimension in our models introduces a number of new phenomena, like

(A) Arnold diffusion,
(B) Complex instability, and
(C) Collisions of bifurcations.

Arnold diffusion is in many respects similar to diffusion through cantori. In a 3-D conservative system, the motion takes place on a 5-D manifold, while the KAM tori are three-dimensional; by subtracting two dimensions, we find one-dimensional tori (lines) in a three-dimensional space. Therefore, we can visualize Arnold diffusion by considering the three-dimensional motion of a particle in a room of perpendicular strings. The set of strings may have a large measure, but the strings do not form continuous surfaces. Thus, there are openings permitting a diffusion in the same way as the holes of the cantori.

It is known that Arnold diffusion is an extremely slow process.[23] In a problem considered by Contopoulos, Galgani, and Giorgilli,[24] two stochastic regions were found that did not communicate with each other even after more than 10^5 periods. Similar results were found by Pettini and Vulpiani.[25] Therefore, Arnold diffusion is probably quite unimportant in galactic dynamics. On the other hand, in the solar system and in plasmas, it may be significant. However, a more detailed study is needed in these cases.

Complex instability is a new type of instability of periodic orbits that occurs only in systems of three or more degrees of freedom. It appears when two pairs of eigenvalues collide on the unit circle and go out of this circle in the complex plane. In systems of two degrees of freedom, a family becomes unstable when two eigenvalues become equal to $+1$ or -1 and then, as the energy increases, move along the real axis. At the same time, a new family bifurcates from the original family, and it is stable if it exists for energies larger than the energy at bifurcation. On the other hand, complex instability is not followed by the bifurcation of another family of periodic orbits. Thus, it introduces stochasticity abruptly and not through an infinite cascade of bifurcations of the Feigenbaum[26] type. We started a detailed study of the stochasticity in complex unstable regions in collaboration with B. Barbanis. We found that a simple periodic orbit in the Hamiltonian

$$H = \tfrac{1}{2}(\dot{x}^2 + \dot{y}^2 + \dot{z}^2 + Ax^2 + By^2 + Cz^2) - \epsilon x z^2 - \eta y z^2 \tag{4}$$

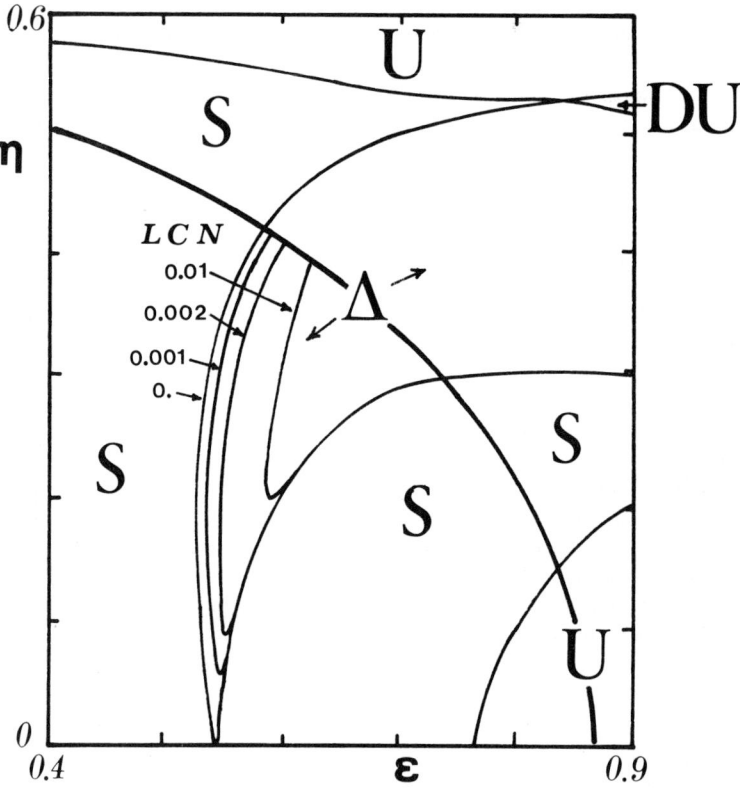

FIGURE 11. A complex unstable region in the parameter space (ϵ, η) for the family $1a$. The letters S, U, DU, and Δ mean stable, simply unstable, doubly unstable, and complex unstable. The thick line corresponds to the escape energy. Orbits with (ϵ, η) inside it cannot escape, while orbits outside it may escape. We also mark approximately the lines of equal maximal Lyapunov characteristic numbers (LCN).

is complex unstable in a certain region of the parameter space (ϵ, η) (while the value of the energy, h, is kept constant; FIGURE 11). The stability types of this family, called $1a$, have been studied by Contopoulos and Magnenat.[27] The family may be stable, simply unstable, doubly unstable, or complex unstable. The complex unstable region joins a stable or a doubly unstable region, except at particular points where it is tangent to a simply unstable region. The complex unstable region reaches the ϵ-axis at only one point because, for $\eta = 0$, the problem is effectively two-dimensional and cannot have complex instability.

We calculated the maximal Lyapunov characteristic number (LCN) for several orbits and for a number of (ϵ, η) values inside the complex region. In FIGURE 11, we mark, in a rough approximation, the lines of equal values of LCN.

The LCN is zero in the stable regions. As ϵ increases for fixed η, the LCN increases slowly as we enter the complex unstable region and then drops rather abruptly to zero. This behavior explains why orbits close to the left boundary of the region Δ show small stochasticity, while orbits close to the right boundary show a much larger stochasticity.[28]

Beyond the escape line of FIGURE 11, all nonperiodic orbits in the Δ region were found to escape after a sufficiently long time. We conclude that complex instability is important in producing stochasticity and escapes in a dynamical system.

REFERENCES

1. GREENE, J. M. 1979. J. Math Phys. **20**: 1183.
2. SHENKER, S. J. & L. P. KADANOFF. 1982. J. Stat. Phys. **27**: 631.
3. MACKAY, R. S. 1982. Ph.D. thesis. Princeton University.
4. AUBRY, S. 1978. Solitons and Condensed Matter Physics. A. R. Bishop & T. Schneider, Eds.: 264. Springer-Verlag. Heidelberg; AUBRY, S. & P. Y. LE DOERON. 1983. Physica **8D**: 381.
5. PERCIVAL, I. C. 1979. Nonlinear Dynamics and the Beam-Beam Interaction. M. Month & J. C. Herrera, Eds.: 302. A.I.P. Conf. Proc. **57**.
6. MATHER, J. N. 1982. Topology **21**: 457.
7. KATOK, A. 1982. Ergodic Theory Dyn. Syst. **2**: 185.
8. BENSIMON, D. & L. P. KADANOFF. 1984. Physica **13D**: 82.
9. MACKAY, R. S., J. D. MEISS & I. C. PERCIVAL. 1984. Physica **13D**: 55.
10. CONTOPOULOS, G. 1971. Astron. J. **76**: 147.
11. SHIRTS, R. B. & W. P. REINHARDT. J. Chem. Phys. **77**: 5204.
12. MENJUK, C. R. 1983. Phys. Fluids **26**: 705.
13. KARNEY, C. F. F. 1983. Physica **8D**: 360.
14. SCHMIDT, G. & J. BIALEK. 1982. Physica **5D**: 397.
15. CONTOPOULOS, G. & M. MOUTSOULAS. 1985. Astron. J. **70**: 817.
16. CHURCHILL, R. G., G. PECELLI & D. L. ROD. 1979. Stochastic Behaviour in Classical and Quantum Hamiltonian Systems. G. Casati & J. Ford, Eds.: 76. Springer-Verlag. Heidelberg.
17. CONTOPOULOS, G. & M. ZIKIDES. 1980. Astron. Astrophys. **90**: 198.
18. CONTOPOULOS, G., CH. VARVOGLIS & B. BARBANIS. 1987. Astron. Astrophys. **172**: 55.
19. PETROU, M. 1984. Mon. Not. R. Astron. Soc. **211**: 283.
20. CONTOPOULOS, G. 1981. Celest. Mech. **24**: 355.
21. BARBANIS, B. 1985. Celest. Mech. **36**: 257.
22. BARBANIS, B. 1987. Celest. Mech. In press.
23. NEHOROSHEV, N. N. 1977. Russ. Math. Surv. **32**: 1.
24. CONTOPOULOS, G., L. GALGANI & A. GIORGILLI. 1978. Phys. Rev. **A18**: 786.
25. PETTINI, M. & A. VULPIANI. 1984. Phys. Lett. **106A**: 207.
26. FEIGENBAUM, M. J. 1978. J. Stat. Phys. **19**: 25.
27. CONTOPOULOS, G. & P. MAGNENAT. 1985. Celest. Mech. **37**: 387.
28. MAGNENAT, P. 1982. Astron. Astrophys. **108**: 89.

Galactic Models with Moderate Stochasticity

MARTIN SCHWARZSCHILD

Princeton University Observatory
Princeton, New Jersey 08544

G. Contopoulos has just described to us the particular type of chaotic behavior that seems to be the dominant one in galactic potentials. In contrast, I would like to walk with you through a sequence of potentials relevant for galaxies—starting with cases totally free of chaos, proceeding in five cautious steps through cases only mildly afflicted by chaos, and ending with cases containing much chaos, like those discussed by G. Contopoulos, or even worse.

SEPARABLE POTENTIALS

The existence of potentials separable in ellipsoidal coordinates has long been known. However, only recently has T. de Zeeuw[1] developed the theory of these Stäckel potentials to the point of astronomical usefulness. The "perfect ellipsoid" corresponds to one of the Stäckel potentials and has been studied in detail. It has a phase-space structure that (except for its total lack of chaos) appears to be typical for triaxial galactic systems.

FIGURE 1 (from T. de Zeeuw[1]) shows schematically the phase-space structure of a perfect ellipsoid. It represents a section of constant energy across action space. The points of the triangle correspond to the three stable periodic orbits. The margin of the triangle is occupied by orbits for which one action is zero. The interior of the triangle is divided into four sections by two dividing lines on which all orbits are unstable. Each of the four sections is totally filled with stable regular orbits of one major family. Together, the four major orbit families provide the backbone for the dynamical structure of the perfect ellipsoids, as well as for a wide class of galactic potentials.

The total occupation of action space by regular orbits insures that self-consistent dynamical solutions for perfect ellipsoids exist for all axis ratios, as is shown by T. Statler's[2] numerical constructions. In fact, this work, as well as the analytical work of T. de Zeeuw *et al.*[3] for perfect elliptical disks, shows that a given density distribution of this type is self-consistent not just with one unique dynamical solution, but with a whole set of such solutions. This nonuniqueness is a direct consequence of the simultaneous existence of more than one major orbit family.

POTENTIALS FOR ELLIPTICAL GALAXIES

In spiral galaxies, like our own, a thin disk is a major component for which a two-dimensional axisymmetric model is a reasonable first approximation. In contrast, elliptical galaxies seem to be genuinely three-dimensional and recent observations even suggest that they are generally not axisymmetric. Accordingly, elliptical galaxies (as well as the bulges in the middle of spiral galaxies) have lately been modeled by triaxial figures (with reflection symmetry with respect to each of three perpendicular planes) having slow or zero figure rotation. For the density distribution in these figures, one of two profiles has generally been used: the modified Hubble profile ($\rho \propto r^{-3}$ for large r) that represents the light profile of an elliptical galaxy in first approximation, or that

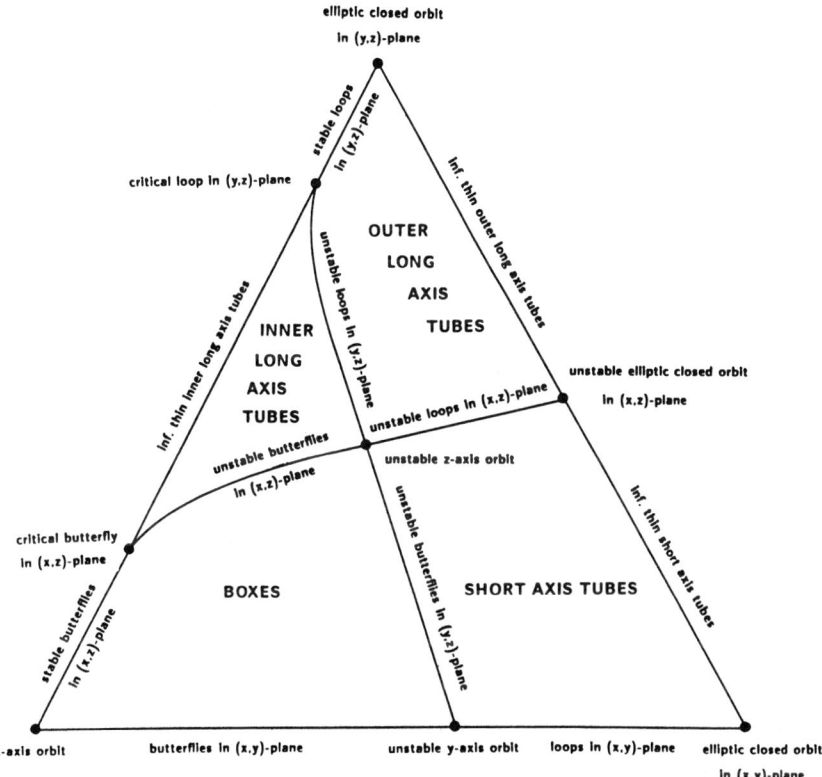

FIGURE 1. Section through action space for a perfect ellipsoid. The four major orbit families occupy the section completely, leaving no space for stochastic orbits.

density profile ($\rho \propto r^{-2}$ at large r) that corresponds to a logarithmic potential and that appears to represent the total mass distribution in such a galaxy fairly well.

In triaxial potentials relevant to elliptical galaxies, numerical experiments with both these profiles have shown that the majority of the orbits are regular, that is, have three effective integrals. The same experiments have also shown that these potentials have a structure in phase-space, or action space, much the same as that of a triaxial

Stäckel potential. Specifically, the bulk of the orbits belong to the same four major orbit families.

These galactic potentials differ, however, in one significant point from separable Stäckel potentials. For the former, a fraction of phase-space is occupied by stochastic orbits, while in the latter, no stochastic orbits exist. In nonseparable galactic potentials, stochastic orbits occur even if the figure rotation is zero and even in the axisymmetric limit.

An example of such stochastic orbits is shown in FIGURE 2 (from Goodman and Schwarzschild[4]). Only box orbits are included in this graph, which therefore corre-

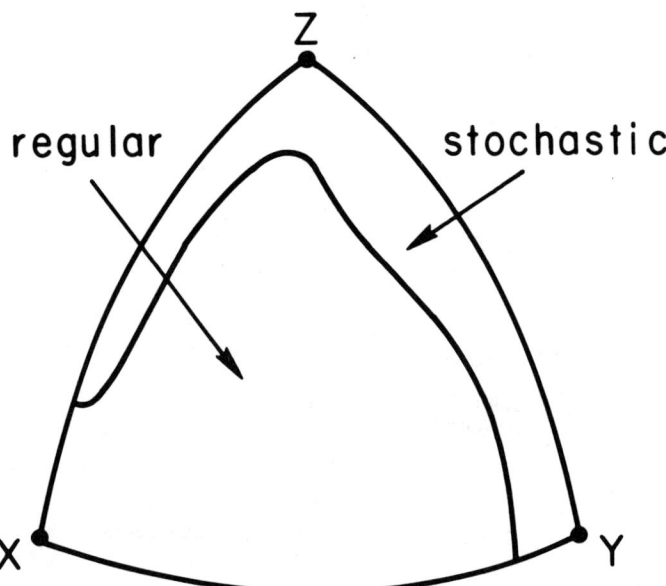

FIGURE 2. Octant of turning point space for box orbits in a galactic potential with a modified Hubble profile. The major portion of the octant is occupied by regular orbits, but a smaller portion, adjacent to the intermediate axis (y) orbit and the short axis (z) orbit, contains stochastic orbits.

sponds only to the lower left quarter of FIGURE 1. Each box orbit is represented in FIGURE 2 by a point according to the Cartesian coordinates of its triple turning point. The three corners of the octant shown in the figure represent the axial orbits, of which only the long axis (x) orbit is stable. The octant is divided into two areas; the larger one is occupied by regular box orbits, but the smaller one is occupied by stochastic orbits. The narrow area of the stochastic orbits stretches along those margins that correspond exactly to the margins of the box orbits in FIGURE 1 (Stäckel potential) that are occupied by unstable orbits.

These numerical experiments indicate not only the narrowness of the stochastic region for galactic potentials, but also one further restraint: the orbits of this region are not honestly stochastic, in the following sense. For an interval at least as long as a Hubble time (about a thousand dynamical times), such an orbit does not wander about all over that part of phase-space allowed by its energy and unoccupied by regular orbits. Instead, it remains in a small section of the allowed phase-space, though not in a strictly three-dimensional subspace as regular orbits do. It is not known whether this restraint is caused by unidentified Cantori. However, in any case, such restrained stochastic orbits appear less dangerous to the existence of self-consistent dynamical equilibrium models than fully stochastic orbits.

In addition to the major stochastic sea between the four major orbit families, there is one other feature in which the nonseparable galactic potentials differ from the separable potentials: the occurrence of minor orbit families (around resonances), each imbedded in a major family and occupying a finite, though generally small, volume of phase-space. These minor families do not seem to affect the structure of a dynamical equilibrium model all that much. Yet, according to Binney and Spergel,[5] they may be of consequence for galactic evolution in phases when the potential varies slowly. When an orbit in such a phase transits between major and minor families, it can change its action values. Thus, in these circumstances, actions are not always adiabatically invariant.

CENTRAL SCATTERER

It is an old story that the perfect billiards table (no pockets and no friction) represents a dynamical system with nothing but regular orbits and that this superb system is degraded to a totally stochastic state by just placing a hard cylinder in the center of the table. This story raises a question for galaxies because many of them have a nucleus at their center consisting of a very small, massive stellar cluster, and others are suspected of having a central massive black hole.

The effects of a central point mass on the orbits in the inner portion of a galaxy have recently been investigated by Gerhard,[5] with the following conclusions. In an axisymmetric galaxy, one component of the angular momentum is an integral; hence, all orbits are tube orbits and the majority of them will never approach the central mass close enough to be effectively scattered. Thus, in an axisymmetric potential, a central scatterer has no significant chaotic effect.

In contrast, in a triaxial potential, self-consistent dynamical equilibrium requires that a substantial fraction of the stars are on box orbits; each of the tube families has orbital densities that, in one projection or the other, are perpendicular to the total model density and, hence, without help from box orbits, cannot reproduce the model. However, a star on a box orbit will in due course come arbitrarily close to the center and will then be significantly scattered. The frequency of such scattering events will be significant only for orbits constrained to the innermost portion of the galaxy.

In the inner part of a galaxy with a central point mass, we may speculate that the stars on box orbits will slowly be scattered into a spherical figure. This in turn will force the tube orbits to conform to an axisymmetric figure, so finally any further scattering of this kind will have no further effects. Could this be an example for a little initial chaos managing to change the system just enough to kill its own effectiveness?

FIGURE ROTATION

Chaos in galaxies is encountered to a significant degree when one considers triaxial systems with substantial figure rotation. Here, we are concerned with figure rotation rates so fast that the corotation point lies only just outside the triaxial component of the total system. Such systems appear to be realized among galaxies, namely, in the fairly common barred galaxies.

No non-axisymmetrical potential relevant for galaxies is known that is integrable with nonzero figure rotation (for a discussion, see Vandervoort[7]). However, I know of no proof that such a potential cannot exist.

G. Contopoulos[8] and his colleagues have carried out extensive investigations of triaxial potentials relevant for galaxies with strong figure rotation. They have concentrated specifically on the complicated set of families of closed orbits that, if stable, provide the parents for the families of regular orbits, and, if unstable, lead to stochastic seas. G. Contopoulos has given us a penetrating review of this subject. Here, I would only like to repeat that for potentials appropriate for galaxies, substantial figure rotation always seems to cause large stochastic seas in the neighborhood of corotation.

Notwithstanding the appearance of such significant chaos, systems containing a rotating bar generally appear to support at least one major family of regular orbits (modified box orbits) that suffices for the construction of a self-consistent dynamical equilibrium model for the bar. Pfenniger[9] has recently derived one such model. On the other hand, the dynamics in a barred galaxy is as yet quite unclear for those components that surround the bar, which are fairly axisymmetric and which cover the chaotic region around corotation. Do these components contain many stars on stochastic orbits? Can such orbits stay restrained to the neighborhood of the equatorial plane (as the observations would require)? Would the occupation of stochastic orbits cause a slow evolution of the system?

GALACTIC SYSTEMS WITH MORE CHAOS

I would like to end with a highly qualitative and surely speculative comment. Of all observed galaxies, a fair fraction seem to be in a perturbed phase caused by an encounter or even a merger with another galaxy. Such perturbations would not seem to belong properly to this workshop because they are so chaotic that they appear not likely to be accessible to a systematic theory of chaos. Of all the other observed galaxies, the majority—particularly, of the giant galaxies—seem to be in a settled state suggesting dynamical equilibrium and seem to have figures that either are axisymmetric or have triaxial symmetry. This circumstance leads to the following questions.

Could it be that all figures of lesser symmetry support an insufficient range of regular orbits, that is, contain too much chaos, to permit self-consisent dynamical equilibrium? Furthermore, could it be that galaxies after their violent relaxation phase go through a slower transitory phase in which dynamical equilibrium is attained by substantial occupation of stochastic orbits restrained by Cantori? The lifetime of such a transitory phase would be governed by the diffusion rate of the stochastic orbits through the Cantori and could be long compared to the dynamical time, but shorter than the Hubble time. Finally, could it be that some such purely gravitational processes could preferentially lead to final equilibrium figures with flat rotation curves (characteristic for all types of major galaxies) irrespective of the mixture of light mass points and dark mass points? Clearly, much needs to be answered in the future.

REFERENCES

1. DE ZEEUW, T. 1985. Mon. Not. R. Astron. Soc. **216:** 273–334.
2. STATLER, T. 1987. Astrophys. J. **321** (October 1).
3. DE ZEEUW, T., C. HUNTER & M. SCHWARZSCHILD. 1987. Astrophys. J. **317** (June 15).
4. GOODMAN, J. & M. SCHWARZSCHILD. 1981. Astrophys. J. **245:** 1087–1093.
5. BINNEY, J. & D. SPERGEL. 1984. Mon. Not. R. Astron. Soc. **206:** 159–177.
6. GERHARD, O. 1986. Mon. Not. R. Astron. Soc. **219:** 373–386.
7. VANDERVOORT, P. 1979. Astrophys. J. **232:** 91–105.
8. CONTOPOULOS, G. & TH. PAPAYANNOPOULOS. 1980. Astron. Astrophys. **92:** 33–46.
9. PFENNIGER, D. 1984. Astron. Astrophys. **141:** 171–188.

The Quadratic Zeeman Effect in Moderately Strong Magnetic Fields[a]

SHANNON L. COFFEY,[b] ANDRÉ DEPRIT,[c]
BRUCE MILLER,[c] AND CAROL A. WILLIAMS[d]

[b]*Naval Research Laboratory*
Washington, District of Columbia 20375

[c]*National Bureau of Standards*
Gaithersburg, Maryland 20899

[d]*Department of Mathematics*
University of South Florida
Tampa, Florida 33620

Ainsi font, font, font
Les petites marionnettes,
Elles font, font, font
Trois p'tits tours
Et puis s'en vont.
—A French nursery rhyme

Over the last several years, celestial mechanics has made a fresh and candid assessment of the discipline's resources and of the conditions facing its research. Out of this examination, there has come agreement on strategies that we believe hold the greatest promise for improving the value of our heritage in mathematical astronomy by moving the research to the front ranks of nonlinear dynamics. Implementing these strategies will require bold strokes of creativity and acts of difficult programming that will unfold in the years ahead.

Our focus is on three strategic priorities:

(i) The first is to strengthen and enhance our core resources—calculus of perturbations by Lie transformations, reduction by normalization, geometric interpretation according to the principles of Morse's Global Theory, etc.

(ii) However, these techniques do not themselves provide the growth opportunities that take full advantage of our resources. Thus, our second strategic priority is to build on traditional expertise by blending our experience in analytical expansions to develop new capabilities in automated algebra.

(iii) Our third strategic priority is to make this a genuine contribution to the physics of nonlinear systems. We intend to do so by highlighting the rims of instability where well-structured phase pictures obtained by accentuating approximate symmetries eventually fail to represent the global physics of a dynamical system.

---

[a]The authors of this paper are listed in alphabetical order.

Among the easiest problems to illustrate the new perspectives in celestial mechanics is the Zeeman effect; it is also one that has attracted more than a passing interest on the part of chemical physicists and astrophysicists in recent years.

Symmetry considerations justify a double reduction. Assuming that the Coulomb field dominates the magnetic field, one replaces the quadratic Zeeman effect by an asymptotic—and formal—approximation that is invariant with respect to the group of Delaunay rotations. Because the original problem is also invariant with respect to the group of rotations about the direction of the magnetic field, so is its Delaunay normalization. This additional symmetry conforms the reduced phase-space as a sphere, which is a critical feature so far overlooked in the literature, yet one we now hold to be characteristic of $SO(2)$-invariant perturbed Keplerian systems.

In the present article, the calculations are not carried out beyond the second power of the Larmor frequency; however, we shall publish elsewhere the details of an expansion executed to the sixth order by means of a programming code in LISP on a special purpose workstation.

In the Zeeman problem, as in the theory of artificial satellites, there is a critical inclination or, equivalently, one might say, a critical eccentricity at which the system undergoes a bifurcation. Above the critical inclination, the circular orbits become unstable; hence, they constitute saddle points in the phase flow of the reduced problem. From any one of them, there emanate two homoclinic orbits that each represent an orbit asymptotic to the circular orbit backward in the past and forward into the future. A closed form expression for such singular solutions would be very helpful; indeed, in conjunction with Melnikov's theorem, it would serve to decide which perturbations added to the Zeeman effect would induce chaotic behavior in the neighborhood of the circular solutions when they are unstable.

SYMMETRIES

The second-order differential equation

$$m\ddot{\mathbf{x}} = \frac{qQ}{\epsilon r^3}\mathbf{x} + \frac{q}{c}\dot{\mathbf{x}} \times \mathbf{B} \tag{1}$$

encapsulates what we call here the Zeeman effect. It determines the motion of a particle with mass m and charge q in the Coulomb field induced by a charge Q in a medium of dielectric constant ϵ. The charges q and Q are of opposite signs. Because the particle's position with respect to the center of the Coulomb field is the vector $\mathbf{x}$, $r = \|\mathbf{x}\|$ stands for the particle's distance to the center. On the central field is imposed a constant magnetic field $\mathbf{B}$. All quantities in equation 1 are expressed in Gaussian units; hence, the divisor c (which stands for the speed of light) in the right-hand member of the equation.

With the magnetic field $\mathbf{B}$ being static and uniform, there exists a unique direction $\mathbf{k}$ fixed in space such that

$$\mathbf{B} = B\mathbf{k} \quad \text{and} \quad qB > 0.$$

In accordance with the latter condition, note that $\mathbf{k}$ and $\mathbf{B}$ are oriented the same way when q is > 0, but they are of opposite orientation when q is < 0; by virtue of this convention, no special attention is given to the sign of the particle's charge.

In spite of the appearances, equation 1 depends on two parameters only,

$$\mu = \frac{|qQ|}{m\epsilon} \quad \text{and} \quad \omega = \frac{qB}{2mc},$$

which are, respectively, the Keplerian constant of dimension length^3/time^2 and the Larmor frequency of dimension time^{-1}.

On account of Maxwell's laws, $\nabla \cdot \mathbf{B} = 0$, and there exists a vector field $\mathbf{A}(\mathbf{x})$ such that $\mathbf{B} = \nabla \times \mathbf{A}$. One chooses usually

$$\mathbf{A} = \frac{B}{2} \mathbf{k} \times \mathbf{x},$$

thus satisfying the Coulomb gauge condition $\nabla \cdot \mathbf{A} = 0$. Under these specifications, equation 1 turns out to be equivalent to the vector Lagrangian equation derived from the function

$$\mathcal{L} \equiv \mathcal{L}(\mathbf{x}, \dot{\mathbf{x}}) = \frac{1}{2} \|\dot{\mathbf{x}}\|^2 + \omega \mathbf{k} \cdot (\mathbf{x} \times \dot{\mathbf{x}}) + \frac{\mu}{r}. \tag{2}$$

Because the momentum canonically conjugate to the position $\mathbf{x}$ is the vector field

$$\mathbf{X} = \frac{\partial \mathcal{L}}{\partial \dot{\mathbf{x}}} = \dot{\mathbf{x}} + \omega \mathbf{k} \times \mathbf{x},$$

the Legendre transformation $(\mathbf{x}, \dot{\mathbf{x}}) \rightarrow (\mathbf{x}, \mathbf{X})$ associates the Hamiltonian

$$\mathcal{H} \equiv \mathcal{H}(\mathbf{X}, \mathbf{x}) = \frac{1}{2} \|\mathbf{X} - \omega \mathbf{k} \times \mathbf{x}\|^2 - \frac{\mu}{r} \tag{3}$$

with the Langrangian $\mathcal{L}$.

Clearly, the system defined by Hamiltonian $\mathcal{H}$ admits two kinds of dynamical symmetries:

(1) It is invariant with respect to the group $SO(2)$ of rotations about the center of the Coulomb field around the axis $\mathbf{k}$ of the magnetic field;
(2) Also, it is left unchanged by the canonical transformation $(\mathbf{x}, \mathbf{X}) \rightarrow (-\mathbf{x}, -\mathbf{X})$, which is the extension of the coordinate transformation by inversion through the center of the Coulomb field.

On the one hand, according to Noether's theorem, the symmetries by rotation beget the integral $\mathbf{k} \cdot (\mathbf{x} \times \mathbf{X})$, which is the projection of the particle's angular momentum on the field lines. In generic terms, the Zeeman effect belongs to the class of conservative dynamical systems with two degrees of freedom. On the other hand, the discrete symmetry by central inversion introduces in the Zeeman effect several features not typical of a conservative system with two degrees of freedom. In particular, it renders the averaged phase flow on the orbital spheres symmetric with respect to the center of the Coulomb field.

THE LINEAR ZEEMAN EFFECT

The Hamiltonian equations derived from equation 3, when they are put in the form

$$\dot{\mathbf{x}} + \omega \mathbf{k} \times \mathbf{x} = \mathbf{X},$$

$$\dot{\mathbf{X}} + \omega \mathbf{k} \times \mathbf{X} = -\left(\frac{\mu}{r^3} + \omega^2\right)\mathbf{x} + \omega^2(\mathbf{x} \cdot \mathbf{k})\mathbf{k},$$

indicate that the momentum $\mathbf{X}$ is the particle's velocity in the precessing frame. They also indicate that the primary effect of the magnetic field is a rotation at the uniform rate $-\omega$ about axis $\mathbf{k}$. A time-dependent canonical transformation removes Larmor's precession.[1-3] By application of Cartan's theorem of the moving frame, namely,

$$d\mathbf{x} = d'\mathbf{x} - \omega(\mathbf{k} \times \mathbf{x})\, dt,$$

where $d'\mathbf{x}$ denotes the position's differential in the precessing frame, the Cartan's 1-form $d\Gamma = \mathbf{X} \cdot d\mathbf{x} - \mathcal{H}dt$ becomes

$$d\Gamma = \mathbf{X} \cdot d'\mathbf{x} - [\mathcal{H} + \omega \mathbf{k} \cdot (\mathbf{x} \times \mathbf{X})]\, dt. \tag{4}$$

Therefore, it follows that the equations of motion in the precessing frame of reference stem from the Hamiltonian

$$\mathcal{H} = \frac{1}{2}\|\mathbf{X}\|^2 - \frac{\mu}{r} + \frac{\omega^2}{2}\|\mathbf{k} \times \mathbf{x}\|^2. \tag{5}$$

When the magnetic field is so small that the term in ω^2 may be omitted from equation 5 (an approximation that is referred to nowadays as the linear Zeeman effect), the dynamical system reduces to a Keplerian system in the precessing frame. This is not to say that the energy shift by the invariant quantity $\omega \mathbf{k} \cdot (\mathbf{x} \times \mathbf{X})$ should be dismissed as a mathematical operation without physical significance, but quite the contrary.

Thanks to gradual advances in measurement techniques, it makes sense at present to study the quadratic Zeeman effect where terms of degree two and higher in ω are retained.

THE QUADRATIC ZEEMAN EFFECT

In the scenery determined by the averaged Hamiltonian on the orbital sphere (what these terms mean will be elucidated in the next sections), two classes of solutions constitute landmarks: (a) the polar orbits, and (b) the equatorial orbits. Both singular classes are most easily reached by treating the quadratic Zeeman effect in cylindrical coordinates.

The decomposition

$$\mathbf{x} = (\mathbf{k} \cdot \mathbf{x})\mathbf{k} + \mathbf{k} \times (\mathbf{x} \times \mathbf{k})$$

introduces the coordinate $z = \mathbf{k} \cdot \mathbf{x}$, along with a unit vector $\mathbf{v}$ and a coordinate $\rho \geq 0$, such that $\mathbf{k} \times \mathbf{x} = \rho \mathbf{v}$. If $\mathbf{u} = \mathbf{v} \times \mathbf{k}$, then $\mathbf{x} = z\mathbf{k} + \rho \mathbf{u}$; furthermore, let λ designate the longitude of $\mathbf{u}$ measured in the magnetic equatorial plane from an arbitrary direction fixed in the precessional frame. Owing to Cartan's theorem of the moving frame,

$$d'\mathbf{x} = \mathbf{k}\, dz + \mathbf{u}\, d\rho + (\mathbf{k} \times \mathbf{x})\, d\lambda.$$

The coordinate transformation $\mathbf{x} \to (z, \rho, \lambda)$ is extended into a gauge-free canonical transformation $(\mathbf{x}, \mathbf{X}) \to (z, \rho, \lambda, Z, P, \Lambda)$ by choosing

$$Z = \mathbf{X} \cdot \mathbf{k}, \qquad P = \mathbf{X} \cdot \mathbf{u}, \qquad \Lambda = \mathbf{k} \cdot (\mathbf{x} \times \mathbf{X}).$$

Physically speaking, Z and P are the components of the particle's velocity respectively along and across the field lines, whereas Λ stands for the projection of the particle's angular momentum along the magnetic field. Expressed in the cylindrical phase variables, the Hamiltonian in equation 5 is the function

$$\mathcal{H} \equiv \mathcal{H}(Z, P, \Lambda, z, \rho, -) = \frac{1}{2}\left(Z^2 + P^2 + \frac{\Lambda^2}{\rho^2}\right) - \frac{\mu}{r} + \frac{\omega^2}{2}\rho^2. \tag{6}$$

As a result of the rotational symmetry, the longitude λ is to be ignored in the Hamiltonian in equation 6, and its conjugate momentum Λ is an integral.

(a) **The polar orbits:** By reason of the equation

$$\dot{\lambda} = \partial \mathcal{H}/\partial \Lambda = \Lambda/\rho^2,$$

the magnetic meridian plane containing the particle remains fixed in the precessing frame if and only if $\Lambda = 0$, which means that the exceptional manifold $\Lambda = 0$ consists of all polar orbits. Such orbits are in effect the solutions of the canonical equations derived from the polar Hamiltonian

$$\mathcal{P} \equiv \mathcal{P}(Z, P, z, \rho) = \frac{1}{2}(Z^2 + P^2) - \frac{\mu}{\sqrt{\rho^2 + z^2}} + \frac{\omega^2}{2}\rho^2. \tag{7}$$

Normalization of the Zeeman effect was first carried out for the singular Hamiltonian $\mathcal{P}$ in parabolic variables in the magnetic meridian.[4-6] Extension of the procedure to the general case when $\Lambda \geq 0$ gives rise to exceedingly complicated manipulations, thereby precluding such an approach toward higher order terms in the approximations. Reverting to a suggestion made long ago,[7] rather than the magnetic meridian as the fundamental reference plane, we adopt the particle's orbital plane, that is to say, the plane through the center of the Coulomb field containing both the position and the velocity of the particle.

(b) **The equatorial orbits:** By virtue of Cauchy's uniqueness theorem, the canonical equations derived from equation 6 admit solutions for which z and Z vanish permanently. These equatorial solutions are equivalent to those of the canonical equations derived from the Hamiltonian

$$\mathcal{E} \equiv \mathcal{E}(P, \Lambda, \rho, -) = \frac{1}{2}\left(P^2 + \frac{\Lambda^2}{\rho^2}\right) - \frac{\mu}{\rho} + \frac{\omega^2}{2}\rho^2.$$

Being separable, Hamiltonian $\mathcal{E}$ could serve as an intermediary for a normalization of the full quadratic Zeeman effect in the neighborhood of the manifold equation orbits.[8]

This is an interesting avenue to explore, especially because it would cover magnetic fields of any strength, weak or strong. However, we did not explore it, the reason being that it should involve elliptic functions, and we do not yet have the tools to conduct a perturbation development in terms of elliptic functions.

For lack of a solution in that direction, one could satisfy oneself with normalizing over a part of the manifold of equatorial orbits. Most remarkable among them are the circular orbits whose radius ρ_c is the positive root of the quartic equation[9]

$$\omega^2 \rho^4 + \mu\rho - \Lambda^2 = 0.$$

We hope to return in a future publication to the problem of normalizing the Zeeman effect in the neighborhood of its circular equatorial orbits.

THE METHOD OF SECULAR PERTURBATIONS

At present, we propose to normalize the Zeeman effect under the assumption that the Larmor frequency is small enough that the general Hamiltonian in equation 5 may be regarded as a perturbed Keplerian system. With ω being taken as a small parameter, equation 5 is therefore decomposed as the sum $\mathcal{H} = \mathcal{H}_0 + \omega^2 \mathcal{H}_1$, with the principal term being the Keplerian

$$\mathcal{H}_0 \equiv \mathcal{H}_0(\mathbf{X}, \mathbf{x}) = \frac{1}{2}\|\mathbf{X}\|^2 - \frac{\mu}{r},$$

while the perturbation is the diamagnetic term

$$\mathcal{H}_1 \equiv \mathcal{H}_1(\mathbf{x}) = \tfrac{1}{2}\|\mathbf{k} \times \mathbf{x}\|^2 = \tfrac{1}{2}\rho^2.$$

The orbital plane, we recall, is the plane through the origin perpendicular to the angular momentum,

$$\mathbf{G} = \mathbf{x} \times \mathbf{X} = G\mathbf{n},$$

with the scalar $G \geq 0$ being the norm of $\mathbf{G}$ and the unit vector $\mathbf{n}$ being the direction normal to the orbital plane. The angle I such that

$$\mathbf{k} \cdot \mathbf{n} = \cos I \quad \text{and} \quad 0 \leq I \leq \pi$$

is the inclination of the orbital plane over the magnetic equatorial plane. The direction $\mathbf{l}$ such that

$$\mathbf{k} \times \mathbf{n} = \mathbf{l} \sin I$$

is what astronomers refer to as the ascending node of the orbital plane on the magnetic equatorial plane. Let ν denote the equatorial longitude of $\mathbf{l}$ reckoned from an arbitrary direction fixed in the precessing frame, and introduce the direction $\mathbf{m} = \mathbf{n} \times \mathbf{l}$. By definition of the orbital plane, position and velocity are linear combinations of the type

$$\mathbf{x} = u\mathbf{l} + v\mathbf{m}, \qquad \mathbf{X} = U\mathbf{l} + V\mathbf{m}.$$

According to Cartan's theorem of the moving frame, the variation of the particle's position in the precessing frame is given by the expression

$$d'\mathbf{x} = d''\mathbf{x} + (\mathbf{k}\,dv + \mathbf{l}\,dI) \times \mathbf{x},$$

where $d''\mathbf{x} = \mathbf{l}\,du + \mathbf{m}\,dv$ denotes the position's differential in the frame $(\mathbf{l}, \mathbf{m}, \mathbf{n})$. The differential identity

$$\mathbf{X} \cdot d'\mathbf{x} = \mathbf{X} \cdot d''\mathbf{x} + (\mathbf{G} \cdot \mathbf{k})\,dv$$

shows that by choosing

$$U = \mathbf{X} \cdot \mathbf{l}, \qquad V = \mathbf{X} \cdot \mathbf{m}, \qquad N(=\Lambda) = \mathbf{G} \cdot \mathbf{k}$$

as the momenta conjugate to the coordinates u, v, ν, respectively, one defines a gauge-free canonical transformation $(\mathbf{x}, \mathbf{X}) \to (u, v, \nu, U, V, N)$. In those variables, $r = \sqrt{u^2 + v^2}$ and

$$\mathcal{H}_0 \equiv \mathcal{H}_0(U, V, -, u, v, -) = \tfrac{1}{2}(U^2 + V^2) - \frac{\mu}{r},$$

$$\mathcal{H}_1 \equiv \mathcal{H}_1(U, V, N, u, v, -) = \tfrac{1}{2}(u^2 + v^2 \cos^2 I).$$

It may seem attractive at this point to follow the beaten track of celestial mechanics. A normalization in the present problem is a canonical transformation,

$$(u, v, \lambda, U, V, \Lambda; \omega) \to (u', v', \lambda', U', V', \Lambda'),$$

close to the identity, that is to be expanded in the powers of the small parameter ω. It purports to convert $\mathcal{H}$ into a series

$$\mathcal{H}' = \mathcal{H}'_0 + \omega^2 \mathcal{H}'_1 + \frac{1}{2!}\omega^4 \mathcal{H}'_2 + \cdots$$

in the powers of ω^2 such that the Poisson bracket $(\mathcal{H}'_0; \mathcal{H}')$ vanishes identically. The latter condition implies that $\mathcal{H}'_0$ is, formally speaking, an integral of the normalized system. The additional integral $\mathcal{H}'_0$ serves to reduce the Zeeman effect to a dynamical system with two degrees of freedom. The major difficulty in performing the normalization resides in determining the kernel of the Lie derivative $\mathcal{L}_0 : F \to (F; \mathcal{H}_0)$, that is, the set of functions F such that $\mathcal{L}_0(F) = 0$. In the XIXth century, astronomers found a way of expressing the Lie derivative $\mathcal{L}_0$ associated with a Keplerian system as a single partial derivative. This is accomplished by transformations of variables.

To begin with, the Cartesian coordinates and velocity components in the orbital plane are replaced by their polar equivalents. If θ is the longitude of the particle in the orbital plane reckoned from the ascending node, then

$$u = r\cos\theta \qquad \text{and} \qquad v = r\sin\theta.$$

Adopting for the momenta the quantities R and Θ such that

$$U = R\cos\theta - \frac{\Theta}{r}\sin\theta \qquad \text{and} \qquad V = R\sin\theta + \frac{\Theta}{r}\cos\theta,$$

one arrives at a gauge-free canonical transformation

$$(U, V, u, v) \to (R, \Theta, r, \theta).$$

In the polar coordinates,

$$\mathcal{H}_0 \equiv \mathcal{H}_0(R, \Theta, -, r, -, -) = \frac{1}{2}\left(R^2 + \frac{\Theta^2}{r^2}\right) - \frac{\mu}{r},$$

$$\mathcal{H}_1 \equiv \mathcal{H}_1(-, \Theta, N, r, \theta, -) = \frac{1}{4} r^2 (1 + \cos^2 I + \sin^2 I \cos 2\theta).$$

So far, no restriction has been imposed on the sign of $\mathcal{H}_0$. However, from here on, it will be assumed that $\mathcal{H}_0$ is <0 or that the normalization is confined to the manifold of bound orbits. In this case, further reduction of the Keplerian $\mathcal{H}_0$ is accomplished by resorting to the Delaunay transformation

$$(R, \Theta, N, r, \theta, \nu) \to (L, G, H, l, g, h).$$

It changes the Keplerian main part into the function

$$\mathcal{H}_0 = -\frac{\mu^2}{2L^2},$$

thus producing the Lie derivative as the single partial derivative

$$\mathcal{L}_0 = \frac{\mu^2}{L^3} \frac{\partial}{\partial l}.$$

With f and g standing respectively for the true anomaly and for the argument of perigee, the diamagnetic term turns out to be the function

$$\mathcal{H}_1 \equiv \mathcal{H}_1(L, G, H, l, g, -) = \frac{1}{4} r^2 [1 + \cos^2 I + \sin^2 I \cos(2f + 2g)].$$

The perturbation belongs to the algebra $\mathcal{F}$ of Fourier series in the mean anomaly; hence, the kernel of $\mathcal{L}_0$ consists of the subalgebra of functions that do not depend on l, and the normalization that we have in view is a Delaunay normalization[10,11]

$$(l, g, \nu, L, G, N) \to (l', g', \nu', L', G', N')$$

that converts $\mathcal{H}$ into its average over the mean anomaly. At the first order,

$$\mathcal{H}'_1 = \langle \mathcal{H}_1 \rangle_l = \frac{1}{2\pi} \int_0^{2\pi} \mathcal{H}_1 \, dl$$

$$= \frac{a^2}{8\pi} \left[(1 + \cos^2 I) \int_0^{2\pi} r^2 \, dl + \sin^2 I \int_0^{2\pi} r^2 \cos 2(f + g) \, dl \right].$$

An easy calculation based on properties of the Delaunay transformation supplies the averages of r^2 and $r^2 \cos(2f + 2g)$ over the mean anomaly l. With the semimajor axis $a = L^2/\mu$ and the eccentric anomaly defined by Kepler's equation $E - e \sin E = l$, it is readily seen that

$$r = a(1 - e \cos E), \quad r \cos f = a(\cos E - e), \quad r \sin f = a\sqrt{1 - e^2} \sin E.$$

Then, it follows from Kepler's equation that

$$\partial E/\partial l = a/r,$$

which is a formula we need in order to change the independent variable from the mean anomaly to the eccentric anomaly in the quadratures. Thereafter, straightforward multiplications based on these elementary relations yield the averages

$$\frac{1}{2\pi}\int_0^{2\pi}\left(\frac{r}{a}\right)^2 dl = \frac{1}{2\pi}\int_0^{2\pi}\left(\frac{r}{a}\right)^3 dE = 1 + \frac{3}{2}e^2,$$

$$\frac{1}{2\pi}\int_0^{2\pi}\left(\frac{r}{a}\right)^2 \cos 2(f+g)\, dl = \frac{1}{2\pi}\int_0^{2\pi}\left(\frac{r}{a}\right)^3 \cos^2 f\, dE = \frac{5}{2}e^2.$$

Hence, the first-order term in the normalized diamagnetic term:

$$\mathcal{H}'_1 = \tfrac{1}{4} a'^2 [(1 + \cos^2 I')(1 + \tfrac{3}{2} e'^2) + \tfrac{5}{2} e'^2 \cos 2g' \sin^2 I']. \tag{8}$$

From here on, for the sake of simplifying the notations, the primes are dropped from the averaged Delaunay variables.

THE PHASE-SPACE AS A SPHERE

The Delaunay normalization reduces the quadratic Zeeman effect to a system with two degrees of freedom; further reduction by means of the rotational symmetry maps it onto a conservative system with only one degree of freedom. Graphing the phase orbits of such a system amounts to drawing the level contours of its Hamiltonian. Some authors[11-13] have chosen to do so on the cylinders (G, g). Such representations are somewhat misleading because the phase-space of $\mathcal{H}'_1$ is not made of cylinders, but of spheres. Maps of $\mathcal{H}'_1$ in planes (G, g) are local, not global: they fail to represent faithfully the phase-space in the neighborhood of the circular orbits ($e = 0$) and of the equatorial orbits ($\sin I = 0$). Yet, these are precisely the areas in phase-space where significant events take place.

One realizes that the phase-space of a perturbed Keplerian system that is axially symmetric is indeed a sphere for each pair of the integrals (L, H) after one has introduced the state functions

$$\xi_1 = GLe \sin I \cos g, \quad \xi_2 = GLe \sin I \sin g, \quad \xi_3 = G^2 - \tfrac{1}{2}(L^2 + H^2). \tag{9}$$

From a geometric standpoint, considering that

$$\xi_1^2 + \xi_2^2 = G^2 L^2 e^2 \sin^2 I, \tag{10}$$

$$2\xi_3 = G^2 \sin^2 I - L^2 e^2, \tag{11}$$

there readily follows that

$$\xi_1^2 + \xi_2^2 + \xi_3^2 = \tfrac{1}{4}(L^2 - H^2)^2, \tag{12}$$

which means that above each point in the octant $\{(L, H) : 0 \leq H \leq L\}$, the phase-space is a sphere, here denoted $\mathcal{S}(L, H)$, of radius $(L^2 - H^2)/2$. On the other hand, from a physical standpoint, in terms of the angular momentum **G** and the modified Laplace vector

$$A = \frac{L}{\mu}\left(X \times G - \frac{\mu}{r}x\right),$$

the new coordinates may be expressed as follows:

$$\xi_1 = (G \times A) \cdot k,$$

$$\xi_2 = \|G\| (A \cdot k),$$

$$2\xi_3 = \|G \times k\|^2 - \|A\|^2.$$

The inverse transformation $(\xi_1, \xi_2, \xi_3) \rightarrow (G, g)$ is defined by the formulas:

$$G = \sqrt{L^2 + H^2 + 2\xi_3}/\sqrt{2},$$

$$e = \sqrt{L^2 - H^2 - 2\xi_3}/L\sqrt{2}, \qquad \cos g = \xi_1/\sqrt{\xi_1^2 + \xi_2^2},$$

$$\sin I = \sqrt{L^2 - H^2 + 2\xi_3}/G\sqrt{2}, \qquad \sin g = \xi_2/\sqrt{\xi_1^2 + \xi_2^2}.$$

These identities permit us to think of the averaged phase-space in geometric terms. Along the diagonal $H = L$ in the plane (L, H), the spheres $\mathcal{S}(L, L)$ collapse to a point; each of these points represents an averaged circular orbit with radius $a = L^2/\mu$ in the magnetic equatorial plane. At the other extreme, along the axis $H = 0$, the spheres $\mathcal{S}(L, 0)$ contain all of the averaged polar orbits. Throughout the interval $(0 \leq H \leq L)$, the north pole of the sphere $\mathcal{S}(L, H)$, that is, the point with coordinates $\xi_1 = \xi_2 = 0$, $\xi_3 = (L^2 - H^2)/2$, represents the circular orbit $(e = 0)$ whose inclination is such that $\cos I = H/L$, whereas the south pole, that is, the point with coordinates $\xi_1 = \xi_2 = 0$, $\xi_3 = -(L^2 - H^2)/2$, represents the class of elliptic orbits with eccentricity $e = \sqrt{1 - H^2/L^2}$ in the equatorial plane, which, when $H = 0$, constitutes the class of equatorial ejection orbits $(e = 1)$.

CRITICAL INCLINATIONS

The search for equilibria in the quadratic Zeeman effect will proceed in two steps: first, by looking for possible critical points away from the north and south poles of the spheres $\mathcal{S}(L, H)$, and then by examining whether the poles themselves might be equilibria.

In the open domain covered by the local map (G, g), the phase flow is represented by the differential equations:

$$\dot{g} = \frac{\partial \mathcal{H}'_1}{\partial G} = \frac{\omega^2 a^2}{4G}[-3 + 3e^2 - 5\cos^2 I + 5(e^2 - \sin^2 I)\cos 2g], \qquad (13)$$

$$\dot{G} = \frac{\partial \mathcal{H}'_1}{\partial G} = \frac{5\omega^2 a^2}{4} e^2 \sin^2 I \sin 2g. \qquad (14)$$

Over a domain that does not include the south pole, note that G is $\neq 0$. The right-hand member of equation 14 vanishes for $g \equiv 0 \mod \pi/2$. On the one hand, the cases of $g \equiv 0 \mod \pi$ do not give rise to equilibria within the local map. Indeed, for such values, the right-hand member of equation 13 vanishes only if $e = 1$; hence, only if $G = 0$, that is to say, when $\xi_3 = -(L^2 + H^2)/2$. Because $|\xi_3|$ must be $\leq (L^2 - H^2)/2$, this occurs only when $H = 0$, which is, more precisely, at the south pole of the spheres $\mathcal{S}(L, 0)$, thus outside the local map (G, g). On the other hand, in case $g \equiv (\pi/2) \bmod \pi$, the right-hand member of the same equation vanishes when

$$1 - 5 \cos^2 I = e^2$$

or, equivalently, when

$$G^2 = HL\sqrt{5}. \tag{15}$$

Only the strip on the cylinder (G, g) for which $H \leq G \leq L$ corresponds to possible states of the averaged quadratic Zeeman effect. Hence, according to equation 15, there exist equilibria for $g \equiv (\pi/2) \bmod \pi$ only when $H \leq H_c$, where $H_c = L/\sqrt{5}$. Let us then denote by S_1 and S_2 the equilibria for which g is equal to $\pi/2$ and $-\pi/2$, respectively. They represent ellipses with their semimajor axes perpendicular to the lines of nodes, with eccentricity $e = \sqrt{1 - H/H_c}$ and inclination $I = \cos^{-1}(\sqrt{H/5H_c})$. On the sphere $\mathcal{S}(L, H)$, the equilibria belong to the meridian plane $\xi_1 = 0$ where their coordinates are such that

$$\xi_2^2 = 5HH_c(H - H_c)(H - 5H_c) \quad \text{and} \quad \xi_3 = \tfrac{1}{2}(L^2 - H^2) - 5H_c(H_c - H).$$

As these conversion formulas make clear, for any given positive value of the integral L, while H is made to increase from 0 to H_c, the equilibria S_1 and S_2 rise from the south pole to the north pole to coalesce and then disappear simultaneously at the north pole when H passes the cutoff value H_c.

It is time now to address the issue of stability. For the variational equations,

$$\frac{d}{dt} \delta g = A \delta g + B \delta G,$$

$$\frac{d}{dt} \delta G = -C \delta g - A \delta G,$$

related to equations 13 and 14, the coefficients are the functions

$$A \equiv A(G, g) = \frac{\partial^2 \mathcal{H}'}{\partial G \partial g} = -\frac{5}{2} \frac{\omega^2 a^2}{G} (e^2 + c^2 - 1) \sin 2g,$$

$$B \equiv A(G, g) = \frac{\partial^2 \mathcal{H}'}{\partial G^2} = \frac{1}{4} \frac{\omega^2 a^2}{G^2} [-3 + 3e^2 - 5c^2 - 5(1 - e^2 + 3c^2) \cos 2g],$$

$$C \equiv C(G, g) = \frac{\partial^2 \mathcal{H}'}{\partial g^2} = \frac{5}{2} \omega^2 a^2 e^2 s^2 \cos 2g.$$

For the sake of conciseness, we use the abbreviations $s = \sin I$ and $c = \cos I$. At the equilibria, when they exist, the coefficients are

$$A(\sqrt{5LH}, \pm\pi/2) = 0,$$

$$B(\sqrt{5LH}, +\pi/2) = -\frac{36}{5} \frac{\omega^2 a^2}{H_c^2},$$

$$C(\sqrt{5LH}, \pm\pi/2) = -\frac{1}{2}\frac{\omega^2 a^2}{H_c^2}(H_c - H)(5H_c - H).$$

Considering that B does not depend on H (and that it is strictly negative) and considering also that C is <0 when $0 \leq H < H_c$, it follows that $\lambda^2 = -BC$ is <0 in that interval, or that the equilibria S_1 and S_2 are stable. At the cutoff value, both characteristic exponents vanish, and this may indicate that the two families of ellipses with stationary perigees corresponding to S_1 and S_2, respectively, emanate from a Hopf bifurcation at the north pole. We shall now demonstrate that this is indeed the case.

THE CIRCULAR ORBITS

Having explored the open zone covered by the cylindrical map (G, g), we turn to the polar caps. At the north pole, which is occupied by the circular orbits, the argument of perigee is not defined and we must resort to the equations of motion

$$\dot{\xi}_i = (\xi_i; \mathcal{H}'), \quad \text{for } 1 \leq i \leq 3, \tag{16}$$

in the global coordinates. Evaluating these Poisson brackets is a somewhat delicate task. To begin with, on account of equation 10,

$$(1 + c^2)\left(1 + \frac{3}{2}e^2\right) = 1 + c^2 + 3e^2 - \frac{3(\xi_1^2 + \xi_2^2)}{2G^2L^2},$$

so the Hamiltonian in equation 8 may be set up as the function

$$\mathcal{H}'_1 = \frac{1}{4}a^2\left(1 + c^2 + 3e^2 + \frac{\xi_1^2 - 4\xi_2^2}{G^2L^2}\right).$$

On the face of this result, one should plan on calculating the Poisson brackets in the right-hand members of equation 16 as sums of the type

$$\frac{\omega^2 a^2}{4}\left[(\xi_i; c^2) + 3(\xi_i; e^2) + \frac{1}{G^2L^2}(\xi_i; \xi_1^2 - 4\xi_2^2) - 2\frac{\xi_1^2 - 4\xi_2^2}{G^3L^2}(\xi_i; G)\right].$$

Once it is realized that

$$\frac{\partial}{\partial G}GLe\sin I = -\frac{2\xi_3}{Le\sin I},$$

it is readily obtained that

$$(\xi_1; \xi_2) = 2G\xi_3, \quad (\xi_2; \xi_3) = 2G\xi_1, \quad (\xi_3; \xi_1) = 2G\xi_2.$$

Furthermore, it is immediately seen that

$$(\xi_1; G) = -\xi_2, \quad (\xi_2; G) = \xi_1, \quad (\xi_3; G) = 0,$$

whence it becomes clear that

$$(\xi_1; e^2) = 2(1 - e^2)\frac{\xi_2}{G}, \quad (\xi_2; e^2) = -2(1 - e^2)\frac{\xi_1}{G}, \quad (\xi_3; e^2) = 0,$$

and, likewise, that

$$(\xi_1; c^2) = 2c^2 \frac{\xi_2}{G}, \quad (\xi_2; c^2) = -2c^2 \frac{\xi_1}{G}, \quad (\xi_3; c^2) = 0.$$

The final result is the differential system

$$\dot{\xi}_1 = \frac{\omega^2 a^2}{2G} M_1 \xi_2, \quad \dot{\xi}_2 = \frac{\omega^2 a^2}{2G} M_2 \xi_1, \quad \dot{\xi}_3 = 5 \frac{\omega^2 a^2}{GL^2} \xi_1 \xi_2, \tag{17}$$

with the coefficients

$$M_1 \equiv M_1(\xi_1, \xi_2, G) = \frac{5\xi_1^2}{L^2 G^2} - 1 + e^2 + 5c^2,$$

$$M_2 \equiv M_2(\xi_1, \xi_2, G) = \frac{5\xi_2^2}{L^2 G^2} - 4(1 - e^2).$$

According to the third equation, there are no critical points outside the coordinate planes $\xi_2 = 0$ and $\xi_1 = 0$; moreover, on account of the second equation, the only critical points in the coordinate plane $\xi_2 = 0$ are the poles. Indeed, for a point in that plane to be critical, one must have $e^2 = 1$, and hence $G = 0$, which thus implies that $\xi_1 = 0$. On the other hand, in the plane $\xi_1 = 0$, besides the poles, the points at which $1 - 5c^2 = e^2$ (according to the first equation) are critical; as a matter of fact, these are the equilibria we analyzed in the previous section.

We shall now focus our attention on the equilibrium—call it S_0—at the north pole. According to equation 12, variations from solutions of the system of equation 17 are bound by the identity

$$\xi_1 \delta\xi_1 + \xi_2 \delta\xi_2 + \xi_3 \delta\xi_3 = 0,$$

which says, in effect, that the variations belong to planes tangent to the sphere along the solution. With the equilibrium S_0 being located at the north pole, it follows that $\delta\xi_3$ vanishes permanently. Hence, the variational equations at S_0 reduce to the differential system

$$\frac{d}{dt} \delta\xi_1 = \frac{\omega^2 a^2}{2L} \left(5 \frac{H^2}{L^2} - 1\right) \delta\xi_2, \quad \frac{d}{dt} \delta\xi_2 = -\frac{2\omega^2 a^2}{L} \delta\xi_1.$$

Its characteristic exponents are the roots of the quadratic equation

$$\lambda^2 + \frac{\omega^4 a^4}{L^2} \left(5 \frac{H^2}{L^2} - 1\right) = 0.$$

They are real when $0 \leq H < H_c$, purely imaginary when $H > H_c$, and they both vanish at the cutoff value H_c.

Therefore, the collapse of the two families of ellipses with stationary perigees onto the circular orbit at $H = H_c$ is the result of a Hopf bifurcation. The story is quite analogous to what has been found for the critical inclinations in the theory of artificial satellites.[14,15]

CONCLUSIONS

All indications are that the Zeeman effect after reduction, truncated to the second power of the Larmor frequency, is integrable in closed form. We have not reached a definite conclusion in that regard, but we can already say that the phase function, $\xi_2/G = \mathbf{A} \cdot \mathbf{k}$, is an elliptic function of the time. A complete discussion of the solutions is likely to establish that two homoclinic solutions emanate from the equilibrium S_0 when $H < H_c$. Should it be possible to express the homoclinic solutions in closed form, then it would become practical to use Melnikov's theorem in order to predict under what conditions a perturbation added to the Zeeman effect would trigger chaos in the neighborhood of S_0 when H is $< H_c$.

It might be useful to pursue the normalization beyond the second degree in the Larmor frequency. The method of secular perturbations may not be the most expedient way to do it. Past the elementary infinitesimal contact transformation developed in the fourth section of this paper, the normalization by Lie transformation at the third order in ω raised major complications that our LISP software could not overcome. We advocate a simpler approach: a conversion from polar to parabolic coordinates followed by a time transformation that models the Zeeman effect as a perturbed elliptic oscillator rather than a perturbed Keplerian system. Then, by means of a Lissajous transformation, we executed the normalization up to the twelfth degree in ω. These developments will soon be published.

ACKNOWLEDGMENTS

Research on the quadratic Zeeman effect was undertaken at the suggestion of William Reinhardt of the University of Pennsylvania. It was carried out in the first half of 1985, at a time when S. L. Coffey was a guest worker at the National Bureau of Standards (on sabbatical leave from the Naval Research Laboratory) and when C. A. Williams held a Summer Faculty appointment in the Bureau's Center for Applied Mathematics. The authors are grateful to David Farrelly of the University of California in Los Angeles for useful conversations.

REFERENCES

1. EPSTEIN, P. S. 1916. Zur quantentheorie. Ann. Phys. **51**(4): 168–188.
2. SOMMERFELD, A. 1916. Zur theorie des Zeeman-effekts der Wassenstofflinien, mit einem anhang über den Stark-effekt. Phys. Z. **17**: 491–507.
3. DEBYE, P. 1916. Quantenhypothese und Zeeman-effekt. Phys. Z. **17**: 507–512.
4. REINHARDT, W. P. & D. FARRELLY. 1982. The quadratic Zeeman effect in hydrogen: an example of semi-classical quantization of a strongly non-separable and almost integrable system. J. Phys. **43**(Colloque C2): 29–43.
5. ROBNIK, M. 1984. The algebraic quantisation of the Birkhoff–Gustavson normal form. J. Phys. A: Math. Gen. **17**: 109–130.
6. ROBNIK, M. & E. SCHRÜFER. 1985. Hydrogen atom in a strong field: calculation of the energy levels by quantising the normal form of the regularised Kepler Hamiltonian. J. Phys. A: Math. Gen. **18**: L853–L859.
7. HERZFELD, K. F. 1914. Der Zeeman-effekt in den quantentheorien der Serienspektren. Z. Phys. **15**: 193–198.

8. DELANDE, D. & J. C. GAY. 1981. A possible dynamical symmetry in diamagnetism. Phys. Lett. **A82**: 393–398.
9. GAJEWSKI, R. 1970. Charge motion in superimposed coulomb and magnetic fields. Physica **47**: 575–595.
10. DEPRIT, A. 1981. Delaunay normalizations. Celest. Mech. **26**: 9–21.
11. CUSHMAN, R. 1984. Normal form for Hamiltonian vectorfield with periodic flow. *In* Differential Geometric Methods in Mathematical Physics. S. Sternberg, Ed.: 125–144. Reidel. Dordrecht.
11. DELOS, J. B., S. K. KNUDSON & D. W. NOID. 1983. High Rydberg states of an atom in a strong magnetic field. Phys. Rev. Lett. **50**: 579–582.
12. DELOS, J. B., S. K. KNUDSON & D. W. NOID. 1983. Highly excited states of a hydrogen atom in a strong magnetic field. Phys. Rev. **A28**: 7–21.
13. NOID, D. W., S. K. KNUDSON & J. B. DELOS. 1983. Resonant states of the hydrogen atom in strong magnetic fields. Chem. Phys. Lett. **100**: 367–370.
14. COFFEY, S. L., A. DEPRIT & B. R. MILLER. 1986. The nature of the critical inclinations in artificial satellite theory. *In* Space Dynamics and Celestial Mechanics. K. B. Bhatnagar, Ed.: 39–52. Reidel. Dordrecht.
15. COFFEY, S. L., A. DEPRIT & B. R. MILLER. 1986. The critical inclination in artificial satellite theory. Celest. Mech. In print.

Chaotic Behavior in Variable Stars[a]

J. ROBERT BUCHLER

Department of Physics
University of Florida
Gainesville, Florida 32611

INTRODUCTION

It was long thought that a variable luminosity was the property of only a rather small subset of all stars, although it was immediately recognized that variable stars did not form a homogeneous group. It soon also became apparent to the early observers that the variability could be due to external causes, such as a twin companion or the interaction with the surrounding medium, or that it could be of an intrinsic nature, as Shapley was the first to suggest. Thanks to the great improvements in detection capabilities in the last couple of decades, it has now become apparent that essentially all stars show variability of a varying degree. In particular, the sun, which was regarded as the paradigm of steadiness, is now classified as an intrinsic variable star. Generally speaking, the stars that show large amplitude variations, and that were therefore the first to be classified as variable, are known as the Classical Variable Stars. They are interpreted as undergoing radial oscillations, whereas stars with smaller amplitude variations are believed to be undergoing oscillations of a nonradial nature.[1,2]

In the following review, we shall limit ourselves to the classical radial variable stars, although the dimensional reduction formalism applies to the nonradial pulsators as well. The extensive theoretical work on these stars, which originated with Eddington in 1929, has shown that the driving of the pulsations is due to the interaction of the hydrodynamical motion with the heat flux emanating from the interior hydrogen-burning region. It has also shown that the ionization regions in the atmosphere provide the necessary phase-shifts to drive and sustain the oscillations. This work, which has met with considerable success, has mostly concentrated on the modeling of the so-called regular variable stars, which are the Classical Cepheids, the RR Lyrae stars, and the BL Herculis stars.

The early observers already noted that even the regular variable stars showed some irregularity in the period and in the amplitude. The theorists, however, have given very little attention to a true physical explanation of the erratic behavior other than to invoke some ad hoc mechanisms such as burning flashes or convective mixing. As one moves to the yellow and red giants, the pulsations become increasingly erratic to the point where, for the Irregular Variables, it becomes no longer even possible to define an average period. A very nice summary of the observational situation of irregularities in the behavior of variable stars can be found in the review papers by Perdang.[3,4]

---

[a]This work was supported in part by the National Science Foundation.

A satisfactory understanding and explanation of such erratic behavior is totally lacking, except, perhaps, for the pioneering suggestions of Spiegel over the years. We should mention that some highly simplified models that exhibit erratic temporal behavior have been proposed,[5-8] but they are more suggestive than realistic and lack structural stability with respect to generalization.

It is important to draw a distinction between stochastic and chaotic irregularities. By stochastic, we mean that so many degrees of freedom are involved that we are totally unable or unwilling to make a detailed model of their interaction. Chaotic behavior, on the other hand, involves only a small number of degrees of freedom (a low-dimensional embedding phase-space), and one can therefore hope to obtain a better and more detailed understanding of the dynamics of such stars and to model them. When confronted with the task of determining whether an irregular variability is chaotic, the obvious first step would be an analysis of the observational data, such as has been done in other areas of physics and chemistry, where the experimenter has some control over the accuracy and nature of the data.[9-11] Indeed, some attempts have been made to analyze stellar variability for chaos. Perdang[12] claims to have found a fractal dimension of about 1.5 in the solar spectrum in the mHz region; Voges, Atmanspacher, and Scheingraber (private communication) claim a dimension between 2 and 3 in the X-ray emission of the eclipsing Her X1; and Spiegel (private communication) claims a dimension of about 4.8 in an analysis of the sunspot cycles. Auvergne and Baglin[13] have applied a phase-space reconstruction technique and determination of the dimension of the correlation function to an oscillating white dwarf, GD66, but, disappointingly, they had to conclude that the observational data are insufficiently accurate, as well as of insufficiently long coverage, to arrive at a definite answer. This seems to typify the situation in general. In the case of the variable stars of interest to us here, the observational situation is even worse. Not only do these objects have much longer periods, but they also have suffered the neglect of the professional observers and have been left to the amateur observers with, in general, concomitant larger errors of observation.

At the present time, and surely for some time to come, the evidence that the apparent irregular behavior in stars is indeed chaotic will remain largely circumstantial. That chaotic behavior is very plausible comes from the observation that many fluid experiments exhibit chaotic behavior, and one would be astonished if gravitating fluids were exempt therefrom. In the following, we shall give more specific arguments that are based on recent developments in the theory of dissipative dynamical systems. In the next section, we shall briefly describe the dimensional reduction method and the derivation of amplitude equations (AEs), which will form the basis of the new framework within which to describe stellar pulsations. In the third section, we shall summarize the application of these techniques to RR Lyrae and Cepheid stellar models. Of this, we shall primarily conclude: first, that the AEs give a remarkable agreement with the numerical hydrodynamical models; second, that at most three essential modes are involved in the nonlinear pulsations; and, third, that the dynamical behavior is so weakly nonlinear as to involve only the lowest order nonlinearities in the AEs. Armed with these results, we shall then speculate about irregular behavior in the fourth section. Based on the physical expected conditions in certain stars, we shall present specific examples of mode coupling that give rise to chaotic behavior.

THE DIMENSIONAL REDUCTION

The equations of hydrodynamics and heat flow in a Lagrangian description can be written in the very compact form,[14,15]

$$\frac{dz}{dt} = \mathbf{R}(z) = \mathbf{L}z + \mathbf{N}_2(z,z) + \mathbf{N}_3(z,z,z) + \ldots, \quad (1)$$

where $\mathbf{L}, \mathbf{N}_2, \mathbf{N}_3, \ldots$ denote spatial operators. The quantity z represents the deviation from static equilibrium of the position vector, the velocity, and the specific entropy of the fluid elements. Implicit in equation 1 is the assumption that the oscillations are sufficiently weakly nonlinear, so we are justified in making the expansion. It is worth mentioning that this system has a very general form and that the techniques, which we shall describe, therefore have a wide range of applicability.

In the following, we shall limit our discussion to radial modes of oscillation of the star. For numerical reasons, it is convenient to discretize the star into n spherical constant mass shells. The quantity z then becomes a vector in a $3n$-dimensional space with components

$$z = (\delta R_1, \delta R_2, \ldots, \delta R_n, \delta v_1, \ldots, \delta v_n, \delta s_1, \ldots, \delta s_n), \quad (2)$$

where the δR_i, δv_i, and δs_i denote the deviation from equilibrium of the radius, velocity, and entropy of the i-th zone, respectively. The linear operator $\mathbf{L}$ becomes a $3n \times 3n$ matrix.

The linear operator $\mathbf{L}$ is real, but not self-adjoint, so its (so-called linear nonadiabatic or LNA) eigenvalues ($\sigma_k \equiv i\omega_k + \kappa_k$) are either real or they come in complex conjugate pairs. The eigenvectors are not orthogonal and one therefore needs a dual basis for a spectral representation of the operator $\mathbf{L}$. Introducing an $\exp(\sigma t)$ time-dependence and using Dirac's notation, we have the eigenvalue problem:

$$\mathbf{L}|\alpha\rangle = \sigma_\alpha |\alpha\rangle, \quad (3)$$

$$\langle \alpha | \mathbf{L} = \sigma_\alpha \langle \alpha |. \quad (4)$$

The dual basis is orthogonal and can be normalized to unity.

Except in special situations, the real branch is made up of the thermal modes, which range from the secular modes with small eigenvalues to the diffusion modes with very large eigenvalues. The complex modes are the vibrational modes.

In addition to the assumption of weak nonlinearity, the dimensional reduction technique requires that it be possible to split the modes into two categories—the first, which comprises the "marginal" (unstable and stable) modes, and the second, which is composed of the very stable or "slave" modes. In addition to the assumed small parameters, κ/ω, for the marginal modes, there may be other small parameters in the stellar model, for example, a resonance parameter for a 2:1 resonance, $\Delta\omega = (\omega_\beta - 2\omega_\alpha)/(\omega_\beta + 2\omega_\alpha)$, or a near dynamical instability, $\omega_0 \sim 0$. The form of the set of AEs will depend on what the small parameters are for given physical conditions.

The AEs can be derived in a variety of ways. Buchler and Goupil,[14] for example, have used a multitime perturbation method. Coullet and Spiegel[16] have developed a particularly elegant and general method. We shall not describe here the details of the derivations, which can be found in those papers. The AEs, which are also sometimes called normal form equations,[17] have the general form

$$\frac{da_\alpha}{dt} = \sigma_\alpha a_\alpha + \text{essential nonlinear terms},$$

$$\frac{da_\beta}{dt} = \sigma_\beta a_\beta + \text{essential nonlinear terms},$$

$$\vdots \tag{5}$$

where the indices $\alpha, \beta, \ldots$ run over all the marginal modes. The solution itself is

$$z = \sum_{\text{marginal modes}} a_\alpha(t) \mid \alpha \rangle + \text{higher order nonlinear terms}. \tag{6}$$

By assumption, the amplitudes of the marginal modes can be separated into a rapidly varying part and a slowly varying part, $a(t) = \tilde{a}(t) \exp(i\omega t)$.

The dimensional reduction method gives not just the form of the AEs in each case, but it also gives specific expressions for their nonlinear coupling coefficients in terms of integrals over the stellar structure and products of LNA eigenvectors, which can be computed.[18] Earlier work with AEs in the context of stellar pulsations was based on the use of adiabatic eigenvectors, which, although yielding the correct form of the AEs, had the flaw that the coupling coefficients could not be computed reliably[19-23] and that dissipation was introduced in an ad hoc way.[24]

It is important to stress that the dimensional reduction is not merely a modal truncation with the dynamic taking place in the submanifold spanned by the marginal eigenvectors, but that the slave modes deform this manifold. They also modify the coefficients of the nonlinear terms in the AEs.

Finally, the question arises as to how far from marginality we can push the theory and what modes should be included in the marginal category. This is so because, after all, the real parts of the eigenvalues vary smoothly from small positive to large and negative. It is well known that asymptotic techniques such as this generally have a much wider range of validity, at least in a qualitative sense, than the restriction of their assumptions would indicate. It is the underlying structure and motion of the poles that ultimately determine the range of validity and the occurrence of bifurcations or crisis points. In realistic physical examples, an analysis of the location of the poles is not feasible and, in practice, the range of validity has therefore to be determined empirically.

To summarize, the dimensional reduction technique has reduced the partial differential system of hydrodynamics and heat transfer to a (hopefully) small set of ordinary differential equations that govern the temporal behavior of the amplitudes of the marginal modes. The amplitudes of the marginal modes play the role of generalized coordinates. The essence of the dynamical behavior is then played in a small subspace of phase-space, towards which the system gets attracted on a very short time scale. Because we are after the long-range behavior of the system, these short-lived transients are of no interest and are ignored by the method.

One really expects only a small number of modes to dominate the dynamical behavior in the classical radial pulsators, so the AE formalism might constitute a very useful approach. In the next section, we summarize the application of this formalism to RR Lyrae stars and Cepheids, including bump Cepheids, and show that the agreement is rather remarkable between theory and numerical hydrodynamics, on the one hand, and observation, on the other. For the nonradial pulsators, the situation is perhaps less encouraging and no such application has as yet been performed. First, the LNA problem itself, which is the starting point in our approach, has not been solved satisfactorily. It is very difficult to obtain physically reliable growth-rates as long as we cannot resolve the dissipative structure on the small scale required by the nonradial modes. Second, the spectrum of the nonradial modes is very dense and many modes are generally expected to be involved, thus leading to a large set of coupled AEs.

APPLICATION TO REGULAR PULSATORS

In this section, we shall summarize the applications of the dimensional reduction techniques to the nonlinear behavior of some regular radial pulsators and we shall show that there is every indication that this behavior takes place on a very low dimensional manifold.

First, we shall consider RR Lyrae models. Buchler and Kovács[25] considered the AEs that are appropriate for the nonresonant interaction between the fundamental and first overtone and they truncated these equations at the lowest (the cubic) nonlinearities. The AEs are

$$\frac{da}{dt} = \sigma_0 a + Q_0 |a|^2 a + T_0 |b|^2 a, \tag{7a}$$

$$\frac{db}{dt} = \sigma_1 b + Q_1 |b|^2 b + T_1 |a|^2 b, \tag{7b}$$

where a and b denote the complex amplitudes of the fundamental and first overtone, respectively, and the cubic coefficients are complicated integrals of products of eigenvectors over the structure of the star. They showed that if the real parts of the cubic coefficients are negative, then the fixed points of these AEs could fully describe the different observed modal behavior of RR Lyrae stars. This includes steady, stable single-mode pulsations in the fundamental or in the first overtone and doubly periodic pulsations in both modes simultaneously.

In order to test the hypothesis that two AEs (which, in addition, are truncated at the cubic nonlinearities) give more than a qualitative description, Kovács and Buchler[26,27] decided to test these AEs on the numerically generated behavior of RR Lyrae models that were initiated with some perturbation and that were let to evolve towards their final states. This test is far more stringent because, on the one hand, it is quantitative, and, on the other, it involves not just the fixed points of the AEs, but also the evolution towards the fixed points. They found that the approach to the nonlinear pulsation, after a transient of negligible duration, involves only the linearly unstable modes. In one particular example,[26] the stellar model had a mass of 0.65 and a luminosity of 60 (both in solar untis), an effective temperature of 7000 K, a composition of ($X = 0.700$, $Z = 0.001$), and was linearly unstable in both the fundamental and in the first overtone. The model parameters were chosen so that the model should show stable sustained nonlinear oscillations in either the fundamental or in the first overtone (depending on the initial conditions). The linear frequencies, ν, and growth-rates, κ ($\sigma = i2\pi\nu + \kappa$), are shown in the last two columns of TABLE 1. Their first numerical hydrodynamical time-integration ("run") was initiated with a velocity profile corresponding to the linear fundamental eigenvector with a 10% admixture of the first overtone, and the second run was done with a 2% admixture. Both were done with a 10-km/s surface velocity. Both runs covered several hundred periods of oscillation. TABLE 1 also shows the spectrum of the eight lowest frequencies, f, that they obtained by a Maximum Entropy Method (MEM) applied to the variation of the stellar radius over a subinterval of the total time base. The second column shows their identification, $\bar{f}$, of these frequencies in terms of two basic ones, $\bar{f}_0$ and $\bar{f}_1$, and their combinations. It is clear from a comparison with the linear spectrum in the third column that $\bar{f}_0$ and $\bar{f}_1$ correspond to the fundamental and first overtone. The other

TABLE 1. The Chosen Model Parameters

MEM Spectrum[a]	Identification of the Frequencies	Linear Spectrum[b]			
$f_1 = 1.849$	$\bar{f}_0 = 1.8478$	$\nu_0 = 1.8488$		$\kappa_0 =$	0.0078
$f_2 = 2.478$	$\bar{F}_1 = 2.4787$	$\nu_1 = 2.4814$		$\kappa_1 =$	0.0547
$f_3 = 3.109$	$2\bar{f}_1 - \bar{f}_0 = 3.1096$	$\nu_2 = 3.1094$		$\kappa_2 =$	-0.0134
$f_4 = 3.697$	$2\bar{f}_0 = 3.6956$	$\nu_3 = 3.8220$		$\kappa_3 =$	-0.2884
$f_5 = 4.326$	$\bar{f}_0 + \bar{f}_1 = 4.3265$				
$f_6 = 0.631$	$\bar{f}_1 - \bar{f}_0 = 0.6309$				
$f_7 = 1.254$	$2\bar{f}_0 - \bar{f}_1 = 1.2169$				
$f_8 = 4.957$	$2\bar{f}_1 = 4.9574$				

[a]The MEM spectrum is of the hydrodynamically generated surface radius.
[b]ν = linear frequency; κ = growth-rate.

modes, if present, are so short-lived that they do not appreciably show up. Kovács and Buchler then obtained the temporal behavior of the amplitudes and phases through a technique of successive fits with Fourier sums.[28] The temporal behavior of the amplitudes of reference 26 is reproduced in FIGURES 1 and 2 as dashed lines.

It is this temporal behavior of the basic (marginal) amplitudes that the AEs are designed to handle. When the complex amplitudes, a, are replaced by moduli, A, and phases, ϕ, with $a = A \exp(i\phi)$, in equations 7a and 7b, the results are

$$\frac{dA}{dt} = \kappa_0 A + \mathrm{Re}\, Q_0 A^3 + \mathrm{Re}\, T_0 AB^2, \tag{8a}$$

$$\frac{dB}{dt} = \kappa_1 B + \mathrm{Re}\, Q_1 B^3 + \mathrm{Re}\, T_1 BA^2, \tag{8b}$$

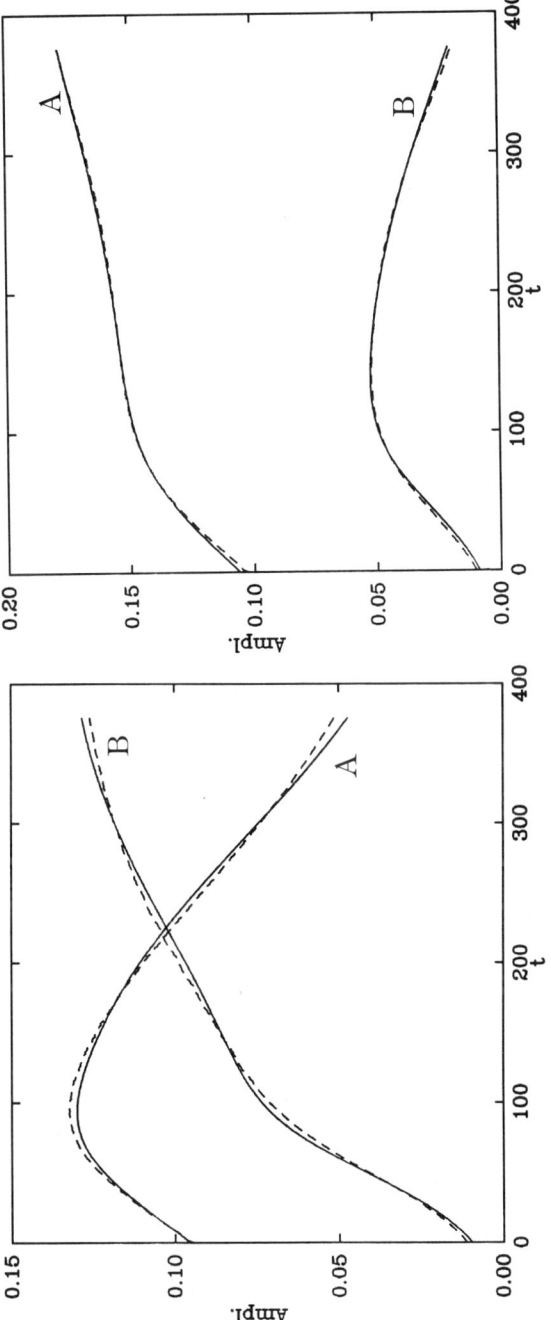

FIGURE 1. Temporal behavior of the amplitudes. Fourier analysis of the data (dashed lines) and fitted integral curves of the AEs (solid lines). The figure on the left is for an initialization with a 10% overtone admixture and the figure on the right is with a 2% overtone admixture.

$$\frac{d\phi_0}{dt} = \omega_0 + \text{Im } Q_0 A^2 + \text{Im } T_0 B^2, \tag{9a}$$

$$\frac{d\phi_1}{dt} = \omega_1 + \text{Im } Q_1 B^2 + \text{Im } T_1 A^2. \tag{9b}$$

The surface radius variation, to lowest order, is given by

$$\frac{\delta R(t)}{R} = \left\{ \frac{1}{2} A(t) e^{i\omega_0 t + i\phi_0} + \frac{1}{2} B(t) e^{i\omega_1 t + i\phi_1} + \text{c.c.} \right\} + \cdots. \tag{10}$$

The normal form equations for the amplitudes and phases are seen to be decoupled.

Taking a phenomenological approach, Buchler and Kovács[26] fitted integral curves of the AEs to the hydrodynamical Fourier amplitudes (the "data") by minimizing the expression

$$S = \sum_{i=1}^{N} \{[A(t_i; \lambda) - A_i]^2 + [B(t_i; \lambda) - B_i]^2\} \tag{11}$$

with respect to the coefficients of the AEs. Here, $A(t; \lambda)$ and $B(t; \lambda)$ represent the integral curves for both initial conditions, and A_i and B_i are the data to be fitted. The integral curves of the AEs with the fitted coefficients are reproduced in FIGURES 1 and 2 as solid lines and they are seen to agree rather well with the hydrodynamical data. The amplitudes of the two limit cycles correspond to the two fixed points of the AEs

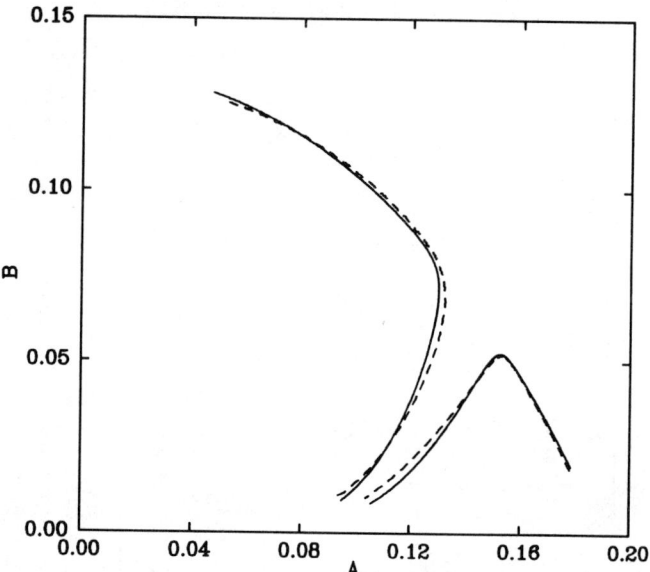

FIGURE 2. Evolution of the same hydrodynamical behavior of the model as in FIGURE 1 in an amplitude-amplitude plot.

TABLE 2. Characteristics of the Limit Cycles of the Model[a]

Integral Fit				Linear Code	
$\kappa_0 = 0.0081$		$\kappa_1 = 0.0560$		$\kappa_0 = 0.0078$	$\kappa_1 = 0.0537$
$\text{Re } Q_0 = -0.222$		$\text{Re } T_0 = -0.864$			
$\text{Re } Q_1 = -3.076$		$\text{Re } T_1 = -2.040$			
$\text{Im } Q_0 = -0.1208$		$\text{Im } T_0 = -0.2791$			
$\text{Im } Q_1 = -0.8429$		$\text{Im } T_1 = -0.8825$			
				Relaxation Method	
$\bar{\kappa}_0 = -0.0185$		$\bar{\kappa}_1 = -0.0076$		$\bar{\kappa}_0 = -0.0185$	$\bar{\kappa}_1 = -0.0050$
$\bar{A}_0 = 0.191$		$\bar{B}_1 = 0.135$		$\bar{A}_0 = 0.191$	$\bar{B}_1 = 0.134$

[a]The left side is the result of the fit, whereas the values on the right side have been obtained from the linear and the relaxation codes, respectively.

(equations 8a and 8b). Their values, as well as their stability coefficients (λ), can be obtained once the coefficients of the AEs are known:

$$A_0^2 = \frac{\kappa_0}{-\text{Re}(Q_0)} \qquad \lambda_1^{(0)} = -2\kappa_0$$
$$\lambda_2^{(0)} = \bar{\kappa}_1 = \kappa_1 + \text{Re}(T_1) A_0^2$$
$$B_1^2 = \frac{\kappa_1}{-\text{Re}(Q_1)} \qquad \lambda_1^{(1)} = -2\kappa_1 \qquad (12)$$
$$\lambda_2^{(1)} = \bar{\kappa}_0 = \kappa_0 + \text{Re}(T_0) B_1^2.$$

The coefficients of the AEs that result from the fit are shown in the first column of TABLE 2 together with the inferred characteristics of the two limit cycles. On the right side, we exhibit the linear growth-rates as obtained with the LNA code. The exact values of the limit cycles and the Floquet exponents were also computed numerically with a numerical hydrodynamical code designed to relax to the periodic solutions. The amplitudes and the real part of the Floquet exponent, which corresponds to the stability coefficients, are shown on the right. The agreement must be considered remarkable, especially because the fit was performed with two hydrodynamical runs that ended long before steady pulsation was achieved.

We therefore conclude that just two AEs, which in addition have been truncated at the lowest (cubic) nonlinearities, provide an almost perfect description of the complicated hydrodynamical evolution of an RR Lyrae model towards its nonlinear pulsational attractors. Because the phases decouple from the amplitudes in this case, the dynamic of the model takes place in a two-dimensional submanifold of phase-space and the most natural generalized coordinates are the moduli of the two amplitudes or, equivalently, the kinetic energies of the two modes.

For a long time, numerical hydrodynamical codes failed to produce sustained doubly periodic nonlinear pulsations,[29] although the AEs in equations 13a and 13b could have double-mode fixed points (for which $A \neq 0$ and $B \neq 0$)[25] when the condition $D = \text{Re}(Q_1)\text{Re}(Q_0) - \text{Re}(T_1)\text{Re}(T_0) > 0$ is fulfilled. In a very recent survey of realistic RR Lyrae models, Kovács and Buchler[27] confirmed that the condition $D > 0$ is never fulfilled in the observed range of stellar parameters. They did however manage to produce sustained double-mode oscillations for helium-poor models ($Y = 0.2$). In their further search, they were guided by the prediction,[30] based on the AEs, that a 2:1 resonance between the fundamental and the third overtone mode could produce conditions favorable to steady, stable doubly periodic pulsations. In order to describe the approach to this (triple-mode) fixed point, it is necessary to add an additional AE for the third overtone and to add the resonant coupling terms, which are quadratic. It is straightforward to show, after modular amplitudes and phases are introduced, that the number of necessary generalized coordinates is four (the three amplitudes and the phase difference, $\phi_3 - 2\phi_0$).

For the purposes of this review, we conclude from this work that the nonlinear hydrodynamical evolution of state-of-the art RR Lyrae models towards their states of steady nonlinear pulsation is rather well described by a very small number (at most four) of coupled nonlinear AEs. In the case of an evolution toward beat pulsation, the presence of the resonance requires the addition of a third AE.

The second application of AEs, which we will now describe here, has been made to bump Cepheids. It has been known for a long time from a Fourier analysis of the light-curves[31] that Cepheids exhibit an abrupt feature in the vicinity of the 10-day period. In 1976, Simon and Schmidt[32] astutely noted on the basis of linear models that a 2:1 resonance between the fundamental and the first overtone seemed to occur concurrently with the observed appearance of a secondary maximum or shoulder ("bump") in the light- and velocity-curves. They therefore conjectured that somehow this resonance is the cause of the bump. Subsequently, a Fourier analysis of the observational data[33,34] showed that the weighted phase difference, $\phi_{21} = \phi_2 - 2\phi_1$, of the first two harmonics has a very characteristic behavior near the resonance, namely, a sharp rise followed by an abrupt drop. This characteristic rise can be understood[18,35] as the nonlinear (quadratic) interaction of two resonant modes. The AEs that are appropriate for such a resonance are given by[14]

$$\frac{da}{dt} = \sigma_0 a + P_0 a^* b \quad \Big| \quad + Q_0 |a|^2 a + T_0 |b|^2 a, \tag{13a}$$

$$\frac{db}{dt} = \sigma_1 b + P_1 a^2 \quad \Big| \quad + Q_1 |b|^2 b + T_1 |a|^2 b, \tag{13b}$$

where, for future reference, we have also indicated the next order cubic interaction terms to the right of the vertical dashed line. When we introduce moduli and phases for the two amplitudes, $a = A \exp(i\phi_\alpha)$ and $b = B \exp(i\phi_\beta)$, the two coupled complex equations (equations 13a and 13b) can be written as three coupled real equations for A, B, and $\Gamma = \phi_\beta - 2\phi_\alpha$.[14]

A reasonable coding effort allows the quadratic coupling terms,

$$P_\alpha = \langle \alpha | \mathbf{N}_2 | \alpha^* \beta + \beta \alpha^* \rangle / \langle \alpha | \alpha \rangle, \tag{14a}$$

$$P_\beta = \langle \beta | \mathbf{N}_2 | \alpha \alpha \rangle / \langle \beta | \beta \rangle, \tag{14b}$$

to be computed directly from the stellar model; Klapp et al.[18] did just that and showed that the predictions of the fixed points of the AEs agreed well with the Fourier analysis of numerical hydrodynamical models to the left of the resonance, but that the AEs failed to give solutions to the right of the resonance. Subsequently, it was shown[35] that the addition of the cubic terms can give a qualitative agreement over the whole range with the numerical hydrodynamical results and with the observational features. Even if the Cepheid stars formed a homogeneous evolutionary sequence, it would not be possible to extract the coefficients of the AEs from the observational data because of the variation of the structure of the stars along that sequence. The fact that the ϕ_{21} phase for the observed Cepheids correlates so well with the period must be considered to be an indication that the Cepheids form a rather homogeneous group.

Before proceeding, we would like to point out that in addition to fixed points, the resonant AEs (equations 13a and 13b) can also have limit cycles that correspond to stable, but periodically modulated oscillations.[14,35] Many RR Lyrae stars exhibit such modulated, but otherwise monoperiodic oscillations (the so-called Blazhko effect). Recently, Moskalik[36] has speculated that the Blazhko effect has its origin in a 2:1 resonance. A detailed numerical hydrodynamical study and its analysis in terms of AEs, such as the one described in the previous section, would be interesting, but it is still lacking.

We conclude that the nonlinear behavior of the regular variables seems to be very well described by the interaction of a small number of modes.

SPECIFIC MECHANISMS FOR IRREGULARITIES

Here, we shall discuss specific types of interaction between a small number of modes that are of astrophysical interest and that give rise to chaotic behavior in some large range of model parameters.

RV Tauri Behavior

RV Tauri stars show alternating high and low minima in their light-curves. Typically, the secondary minimum gradually deepens until the light-curve becomes "W Virginis"–like, that is, with successive minima of the same depth. This alternation generally does not occur regularly.[37]

The AEs for the coupling of two modes that are in a 2:1 resonance (equations 13a and 13b) are well known to give rise to chaotic behavior in some range of parameter values, even when only the quadratic terms are retained in equations 13a and 13b (in the context of plasma physics, see, e.g., reference 38). Here, we just want to give a qualitative idea of what behavior one may expect.

As an example, we have computed the solution of the quadratic AEs in equations 13a and 13b for parameter values chosen to give chaotic behavior ($v = -0.95$, $\Delta\omega = 6.0$, $q = 0.2$ in the notation of reference 35). The orbit lies in the three-dimensional phase-space of A, B, and Γ. Its projection over some limited time interval is shown in FIGURE 3 and it clearly exhibits the irregular cycling of the two amplitudes. The phase Γ (not shown here) also has an irregular behavior. The resultant behavior of the corresponding stellar model in lowest order is given by[14]

$$S(t) \sim A(t) \cos(\omega t) + r B(t) \cos[2\omega t + \Gamma(t)], \qquad (15)$$

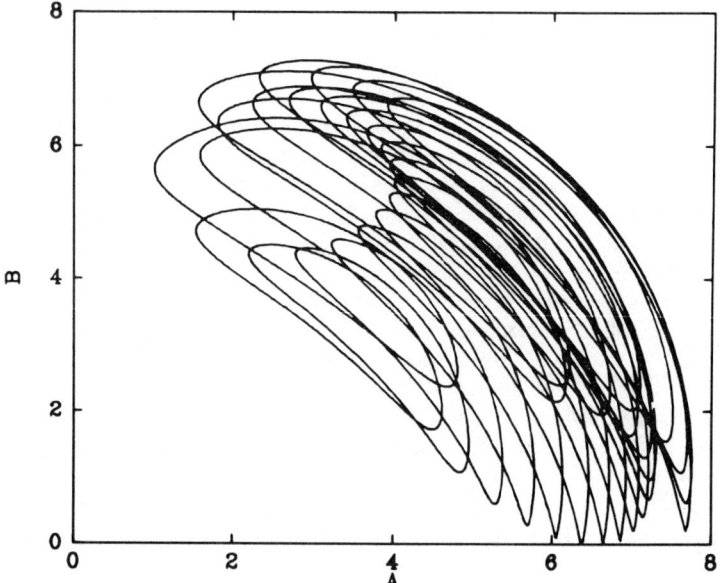

FIGURE 3. Projection of a segment of an integral curve onto the B-A plane.

where r denotes a combination of physical parameters that is different for the light- and the velocity-curves. Under the assumptions inherent in the dimensional reduction method, the oscillatory behavior associated with ω is much more rapid than the variation of the amplitudes and the phase Γ. However, as we mentioned earlier, the range of validity of the formalism is expected to extend far beyond the domain that the assumptions imply.

On the left-hand side of FIGURE 4, we exhibit the observed behavior of the RV Tauri–type star DS Aqr, adapted from Tsesevich.[37] Juxtaposed on the right, we show the shape of the signal $S(t)$ (equation 15) at five equally spaced time intervals on the attractor shown in the previous figure, where we have chosen $r = 3$ and ω much greater than the slow cycling frequency on the attractor. Because the attractor is chaotic, the shape of the signal never repeats exactly. FIGURE 4 shows that typical RV Tauri behavior can be reproduced by resonant 2-mode AEs. This superficial agreement, however, should not be taken too seriously until a detailed study of RV Tauri models has confirmed it. The coefficients of these AEs can be computed directly for a given stellar model as was done for the bump Cepheids.[18] This project is currently under way.

Semiregular and Irregular Stars

Recently, Buchler and Goupil[39] have suggested a mechanism for irregular behavior in red-giant envelopes. It is based on the following physical situation in these stars.

Linear studies[40,41] have shown that these envelopes are near a dynamical instability, that is, that the frequency of oscillation of the fundamental mode of oscillation is small compared to that of the first overtone and is of the same order as the growth- or decay-rates of these modes. If we assume that the temporal behavior of this type of star is dominated by the interaction of these two modes, we obtain the following AEs:

$$\frac{dx}{dt} = y, \tag{16a}$$

$$\frac{dy}{dt} = 2\kappa_0 y + |\sigma_0|^2 x + K_1 x^2 + K_2 |b|^2 + K_3 xy, \tag{16b}$$

$$\frac{db}{dt} = \sigma_1 b + K_4 xb, \tag{16c}$$

where the K's are quadratic coefficients, similar to those of equations 13a and 13b. In lowest order, the stellar radius will exhibit a behavior given by

$$S(t) \sim x(t) + r|b(t)|\cos(\omega_1 t). \tag{17}$$

In FIGURE 5, we have reproduced the projections of a segment of an integral curve of equations 16 onto the y-x and the B-x planes, respectively. The parameters of the AEs were chosen to give a chaotic orbit ($\eta = 3.0, \mu = 6.0, \xi = 1, k = 0.05, L = 0$ in the notation of reference 39). One notes that during part of the cycles, the amplitude b

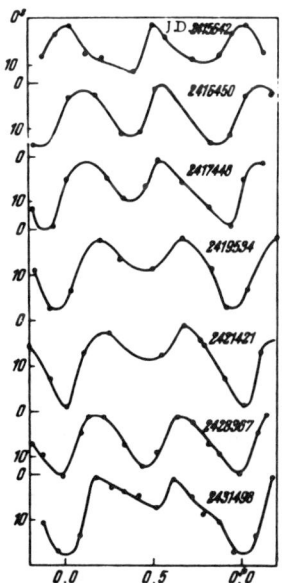
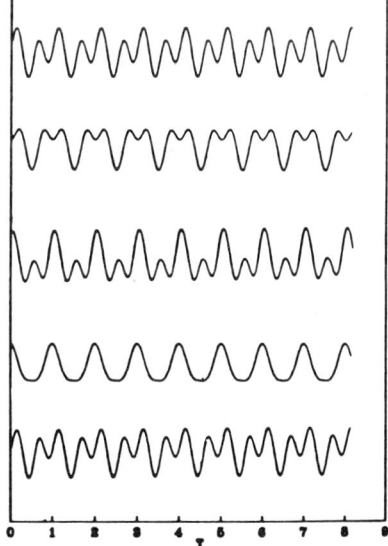

FIGURE 4. Behavior of the luminosity of the star DS Aqr at different epochs (left). Typical shape of the signal at different times as predicted by the AEs (right).

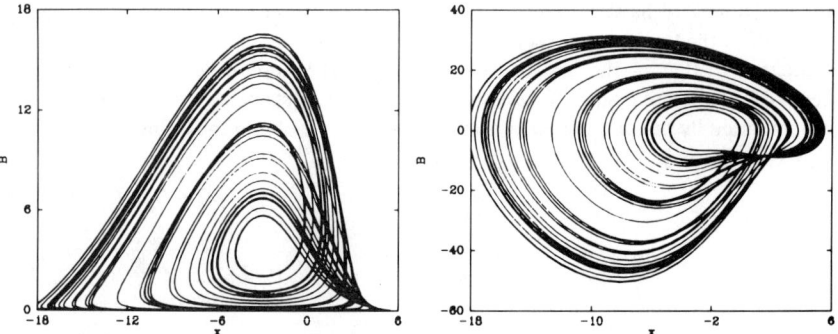

FIGURE 5. Projections of a segment of an integral curve onto the y-x plane (left) and onto the B-x plane (right).

vanishes and, consequently, there are no oscillations on the short time scale (cf. equation 17). Such intermittency is very common in the Irregular Variables. In order to convey a feeling for the possible appearance of the signal, we show the latter for various values of the parameters ω and r in FIGURE 6. For a visual comparison, we also show the appearance of the semiregular variable star μ Cephei[42] in FIGURE 7. Again, we do not claim to have reproduced the behavior of a real star, but we have shown that such characteristic behavior as strong irregularity or intermittency is clearly contained in the modal interaction, which we have just described. A study of realistic models of red-giant envelopes with a mixture of numerical hydrodynamical computations and with AEs should reveal whether our conjecture is right or not.

Spiegel has suggested that massive and supermassive stars are also candidates for irregular behavior because of the presence of a near dynamical instability. In supermassive stars, the dynamical instability is caused by the fact that (a) radiation pressure dominates with a concomitant adiabatic index close to 4/3 and (b) general relativistic effects require an average adiabatic index somewhat greater than 4/3 for stability. Despite the very different physical structure of these objects, the underlying dynamic is very similar to the one that we have just discussed and leads to the same type of AEs, but of course with different coefficients.

Small Irregularities in the Regular Variables

Because even the most regular variables are known to exhibit small erratic fluctuations, it is of interest to search for mode couplings that might be responsible for this unsteady behavior. In the spirit of the dimensional reduction approach, we search again for modes that have small parameters. In the discussion of the LNA spectrum, we only briefly mentioned the real modes, among which the secular modes have eigenvalues that are comparable to or even smaller than the growth-rates of the vibrational modes. Thus, these modes need to be included among the marginal modes. The reason why we believe that we have been able to get away with omitting them so far is as follows: the coupling coefficients between these modes and the vibrational modes are very small in view of the small overlap of the respective eigenvectors; therefore, they can only have a small influence on the oscillatory limit cycles. If we start to look for finer effects, however, we need to consider them.

To be specific, let us envisage a scenario in which a linearly unstable vibrational mode α is coupled to several secular modes, $k = 1, 2, \ldots$. The apposite AEs are then

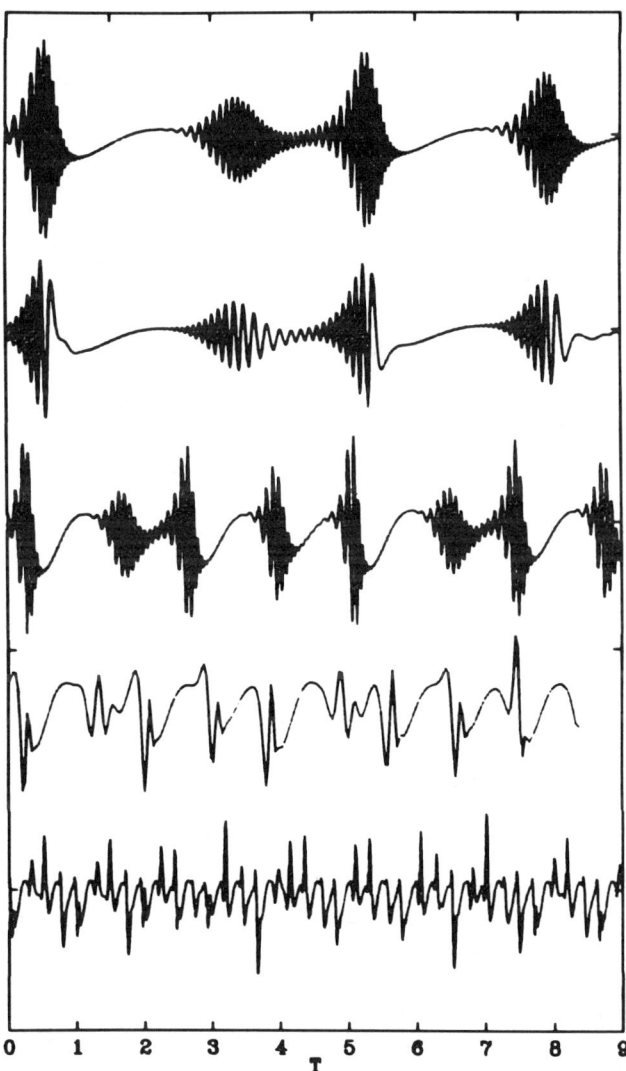

FIGURE 6. Synthesized appearance of the variations for various parameter values.

given by[15]

$$\frac{dA}{dt} = (\kappa_\alpha + R_1 b_1 + R_2 b_2) A + Q_\alpha A^3, \tag{18a}$$

$$\frac{db_k}{dt} = \kappa_k b_k + S_k A^2 + \sum_{j,i} T_{ij} b_i b_j, \tag{18b}$$

where the Latin indices run over all the secular modes that we wish to include. In these equations, we have only kept the cubic coefficient Q_α corresponding to the self-coupling of the vibrational mode. The coefficients R and S are expected to be very small. The

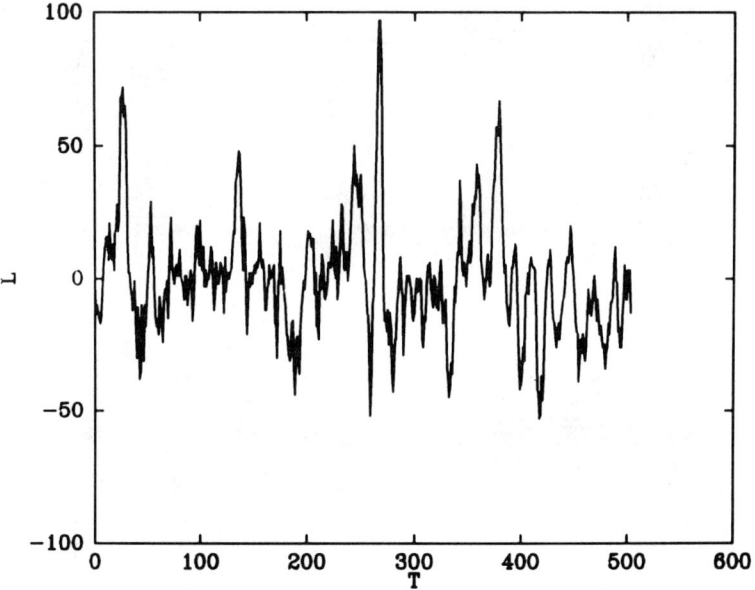

FIGURE 7. Luminosity variations of the semiregular star μ Cephei.

signal, to lowest order, will be of the form

$$S(t) = A(t) \cos(\omega_\alpha t + \phi) + b_1(t) + b_2(t). \tag{19}$$

The fixed point solutions of this system then have the approximate amplitudes:

$$A^2 \sim \frac{\kappa_\alpha}{-\text{Re}(Q_\alpha)}, \quad b_k \sim \frac{S}{\kappa_k} A^2. \tag{20}$$

We have not studied the nature of the solutions of these equations in detail yet, but it is clear that (a) the secular modes can cause a small displacement of the nonlinear pulsation, (b) they can cause a Hopf bifurcation of the fixed point, which, as equation 19 shows, causes (regular) intermittency, and (c) they can give rise to chaotic behavior.

DISCUSSION AND CONCLUSIONS

The dimensional reduction of the original partial differential system of hydrodynamics and heat transfer has several undeniable merits. First, it leads to a physical description of the behavior of the system in terms of the nonlinear interaction between normal modes of oscillation. It is actually the only approach that allows a truly systematical study of the behavior of stellar models in physical terms. Second, it is not just a formal description, but it also works quantitatively remarkably well, at least for the classical radial variable stars to which it has been applied. It has lent weight to the important conjecture that in radial stellar pulsators, there is only a rather small number of modes that dominate the dynamic.

For completeness, I should mention that Perdang[4] has taken a rather different approach to the same problem of irregular behavior in stars. He attacks the problem from a strictly adiabatic point of view. This has the great advantage that the pulsation problem can be cast into a Hamiltonian formulation, for which there exists a number of rigorous theorems. In particular, Perdang shows with numerical experiments on polytropic models that as the number of modes increases, the threshold for stochastic behavior decreases (stochasticity may be considered to be the Hamiltonian analogue of dissipative chaos, although there are important differences). In addition, because the generic Hamiltonian is nonintegrable, he asserts that stochasticity should be very common, especially in nonradial oscillations where the spectrum is much denser and where, generally, many more modes seem to be involved and the energy threshold for stochasticity is low. Real stars, however, are dissipative and are evolving very slowly (on a nuclear time scale), and it is not clear to us at the present time in what sense a Hamiltonian formulation can give a representative description of a dissipative system even in the limit of zero dissipation. For example, in the dissipative chaotic case, the attractor has a fractal structure, whereas in the Hamiltonian limit, it is space filling. In these proceedings, Schmidt addresses some aspects of this question.

We feel that the explanation of the irregular behavior of variable stars poses a very exciting challenge. We are just beginning to get a feeling for the rich underlying physics and to develop the necessary mathematical tools. Clearly, there is a need for more accurate observations and especially for observations that are tailored for this problem. Because the time sequence of any variable can be used to reconstruct the attractor, the brightness variations offer the best hope; this implies, therefore, very accurate photometry and, preferably, equally spaced observations of a given object with some tens of points per fluctuation (e.g., reference 13). At the same time, the interplay between numerical hydrodynamical modeling and the amplitude equation formalism is bound to lead to a better understanding of the observed irregularities of the stars.

REFERENCES

1. LEDOUX, P. & TH. WALRAVEN. 1958. Handb. Phys. **L1:** 353.
2. COX, J. P. 1980. Theory of Stellar Pulsations. Princeton Univ. Press. Princeton, New Jersey.
3. PERDANG, J. 1985. Chaos in Astrophysics. J. R. Buchler, J. Perdang & E. A. Spiegel, Eds.: 11. NATO ASI Ser. **C161**. Reidel. Dordrecht.
4. PERDANG, J. 1985. Physicalia **7:** 239.

5. BAKER, N., D. MOORE & E. A. SPIEGEL. 1966. Astron. J. **71:** 845.
6. AUVERGNE, M., A. BAGLIN & P. J. MOREL. 1981. Astron. Astrophys. **104:** 47.
7. BUCHLER, J. R. & O. REGEV. 1982. Astrophys. J. **263:** 312.
8. WHITNEY, C. A. 1984. *In* Proceedings of the 25th Liège International Astrophys. Colloquium. A. Noëls & M. Gabriel, Eds.: 454. Université de Liège. Cointe-Ougree. Belgium.
9. LIBCHABER, A. & J. MAURER. 1980. J. Phys. **41**(Coll. C3): 51.
10. ROUX, J. C., R. H. SIMOYI & H. L. SWINNEY. 1983. Physica **8D:** 257.
11. ABRAHAM, N. B., J. P. GOLLUB & H. L. SWINNEY. 1984. Physica **11D:** 252.
12. PERDANG, J. 1981. Astrophys. Space Sci. **74:** 149; Mon. Not. R. Astron. Soc. **196:** 109P.
13. AUVERGNE, M. & A. BAGLIN. 1985. Astron. Astrophys. **142:** 388.
14. BUCHLER, J. R. & M. J. GOUPIL. 1984. Astrophys. J. **279:** 394.
15. BUCHLER, J. R. 1985. *In* Chaos in Astrophysics, loc. cit., p. 137.
16. COULLET, P. & E. A. SPIEGEL. 1983. SIAM J. Appl. Math. **43:** 776.
17. GUCKENHEIMER, J. & P. HOLMES. 1986. Nonlinear Oscillations, Dynamical Systems and Bifurcations of Vector Fields. Springer-Verlag. New York/Berlin.
18. KLAPP, J., M. J. GOUPIL & J. R. BUCHLER. 1985. Astrophys. J. **296:** 514.
19. BUCHLER, J. R., W. YUEH & J. PERDANG. 1978. Astrophys. J. **214:** 510.
20. BUCHLER, J. R. & J. PERDANG. 1979. Int. J. Quantum Chem. **13:** 183.
21. VANDAKUROV, YU. V. 1979. Sov. Astron. **23:** 421.
22. DZIEMBOWSKI, W. 1982. Acta Astron. **32:** 147.
23. REGEV, O. & J. R. BUCHLER. 1981. Astrophys. J. **250:** 769; 1982. Astron. Astrophys. **114:** 188.
24. TAKEUTI, M. & T. AIKAWA. 1981. Sci. Rep. Tohoku Univ. (Ser. 8) **2:** no. 3.
25. BUCHLER, J. R. & G. KOVÁCS. 1986. Astrophys. J. **308:** 661.
26. BUCHLER, J. R. & G. KOVÁCS. 1987. Proceedings IAU Symp. 123; Astrophys. J. In press (July 1 issue).
27. KOVÁCS, G. & J. R. BUCHLER. 1987. Astrophys. J. In press; Proceedings IAU Symp. 123.
28. KOVÁCS, G., J. R. BUCHLER & C. G. DAVIS. 1987. Astrophys. J. In press (July 1 issue).
29. COX, A. N. 1980. Annu. Rev. Astron. Astrophys. **18:** 15.
30. DZIEMBOWSKI, W. & G. KOVÁCS. 1984. Mon. Not. R. Astron. Soc. **206:** 497.
31. PAYNE-GAPOSHKIN, C. 1947. Astron. J. **52:** 218.
32. SIMON, N. R. & E. G. SCHMIDT. 1976. Astrophys. J. **205:** 996.
33. SIMON, N. R. & A. S. LEE. 1981. Astrophys. J. **248:** 291.
34. SIMON, N. R. & T. J. TEAYS. 1983. Astrophys. J. **265:** 996.
35. BUCHLER, J. R. & G. KOVÁCS. 1986. Astrophys. J. **303:** 749.
36. MOSKALIK, P. 1986. Acta Astron. Submitted.
37. TSESEVICK, V. P. 1975. *In* Pulsating Stars. B. V. Kukarkin, Ed.: 112. Wiley. New York.
38. WERSINGER, J. M., J. M. FINN & E. OTT. 1980. Phys. Fluids **23:** 1142.
39. BUCHLER, J. R. & M. J. GOUPIL. 1987. Astrophys. Submitted.
40. TUCHMAN, Y., N. SACK & Z. BARKAT. 1978. Astrophys. J. **219:** 183.
41. FOX, M. W. & P. R. WOOD. 1982. Astron. Astrophys. J. **259:** 198.
42. ASHBROOK, J., R. L. DUNCOMBE & A. J. J. VAN WOERKOM. 1954. Astron. J. **59:** 12.

Chaos and the Solar Cycle

E. A. SPIEGEL[a] AND ALAN WOLF[b]

[a]*Astronomy Department*
Columbia University
New York, New York 10027

[b]*Physics Department*
The Cooper Union
New York, New York 10003

Sunspots are the most evident manifestation of the solar cycle and their numbers have been carefully recorded for a century. Serious modeling of this cyclic process starting from physical principles has been attempted in recent decades, but, as hydromagnetic turbulence is clearly involved, no quantitative progress has been made. The activity mirrored in sunspots shows variability in both space and time, as in many turbulent processes, and it is very hard to draw simple conclusions from the data directly. A reduced goal is to try to understand just the global temporal variations in solar activity as revealed in the total number of spots, that is, the Wolf number. The hope has been that a lumped model of this temporal variability might prove of value in understanding the solar cycle.

Investigators in the nineteenth century were struck with the seeming periodicity of the solar cycle. This preoccupation has not changed much, and among the very few mathematical models of the cycle, there were attempts to describe the solar activity as a limit cycle.[1] The inspiration for this kind of model is a representation of the data in a phase plane that showed an oval-shaped phase portrait.[1,2] In this simple picture, the erratic character of the solar cycle is seen as a result of noise generated by physics extraneous to the cycle itself. In this sense, the sunspot cycle would be subject to random or nondeterministic processes. However, it is also possible that the solar cycle is a chaotic process and that most of its character is deterministic. In that case, even the Maunder and other minima[3] might find a natural explanation in chaos theory.[4]

Bracewell[5a] has recently produced a statistical study of the sunspot numbers since 1700. His treatment is aided by a device that he introduced some years ago,[5b] namely, to replace the sunspot number, $M(t)$, by another variable, $N(t)$, with the property that $|N| = M$, and where the sign of N changes at each sunspot minimum. Thus, the 22-year cycle time implied by the magnetic data is imposed. Bracewell has complexified and smoothed the signal M and has plotted his results in a plane. This seems to be the most careful two-dimensional representation of the sunspot data yet displayed. However, as the path of the cycle in this plane crosses itself, Bracewell's plot is not a phase portrait. The crossings are quite definite and do not look as if they are induced by noise. We take his results as an indication that the sunspot dynamics occurs in a state space with greater than two dimensions.

It is not surprising that results like Bracewell's are compatible with the possibility of a nondeterministic solar cycle. Indeed, it may be that the dimension of the process is too high to allow simple modeling. Nevertheless, we feel it is worthwhile to subject the data to an analysis from the standpoint of dynamical system theory in order to help formulate models that may guide us in understanding the qualitative behavior of the solar cycle.

The present work is an examination of the recorded Zurich daily sunspot numbers whose compilation we owe to the kindness of J. A. Eddy. FIGURE 1 shows these data, which are thinned by a factor of two for ease of presentation. The same thinning of the data is used for all the figures. We have constructed phase portraits from these data by the method of delay coordinates.[6] This means that given the sunspot number $M_i =$

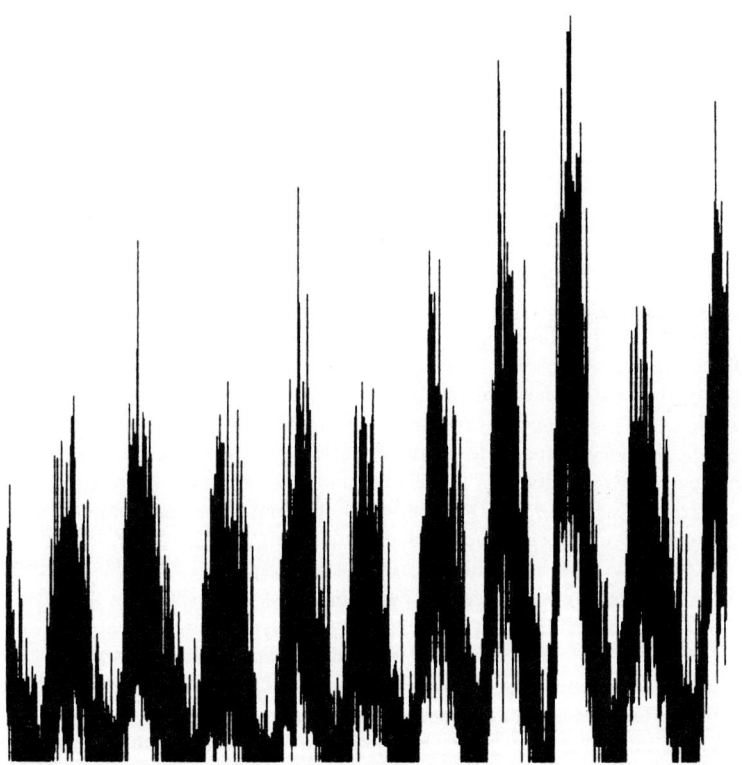

FIGURE 1. The Zurich daily sunspot number, M_i, for 100 years as a function of time. Only the M_i for even i's are shown.

$M(t_i)$, where i is the number of the day of measurement, we constructed triplets of points M_i, M_{i+d}, M_{i+2d}, where the delay time d is an integral number of days. These triplets are plotted in a three-dimensional Euclidean space and are connected sequentially by line segments. We show the phase portrait in FIGURE 2 for the M_i; the figure is projected onto a plane.

After some experimentation, we have adopted $d = 1200$ for FIGURE 2 and the other results shown below. The object in FIGURE 2 has also been rotated for ease of

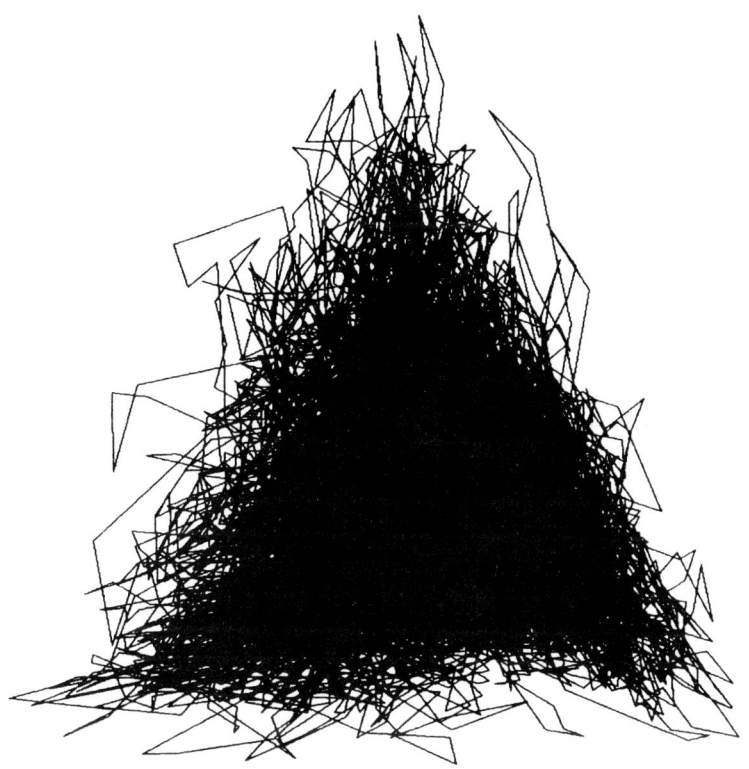

FIGURE 2. A delay reconstruction of the data of FIGURE 1 with a delay of 1200 days.

FIGURE 3. As in FIGURE 2, but here the data have been filtered with a cutoff at 1200 days.

comparison with the other figures; those have been rotated to show an open view of the phase trajectory. To describe the orientation of the phase portraits, we shall speak of x, y, z instead of N_i, N_{i+d}, N_{i+2d}. The object in FIGURE 2 has been rotated 30 degrees about the x-axis, then 60 degrees about the new y-axis.

To get some idea of the phase dynamics, we have filtered the data using a low pass filter.[7] In FIGURE 3, we show a phase portrait for filtered data with the same orientation as for FIGURE 2. The filter has a cutoff of 60 nHz, that is, Fourier components of the sunspot signal with periods less than two days are reduced by 50 db. The drop-off in filter transmission occurs within a transition range of 12 nHz. Our qualitative impression from FIGURE 3 is that there is an interesting dynamical structure. This impression is even stronger when we use the alternated sunspot numbers, N_i. The corresponding delay reconstruction is shown in FIGURE 4.

A striking feature of the reconstructed orbit in FIGURE 4 is the omnipresence of small-scale loops. These represent fluctuations on a time scale of one year. We do not know whether these are an essential aspect of the cycle or not. For the kind of model we are hoping to describe elsewhere, we do not want to cope with this level of detail. Thus, in FIGURE 5, we show the analogue of FIGURE 4 with the filter cutoff lowered to remove the one-year loops superposed on the longer range cyclic variation.

FIGURE 4. The delay reconstruction made from the flipped sunspot numbers, N_i, with the filtering of FIGURE 3 and the orientation described in the text.

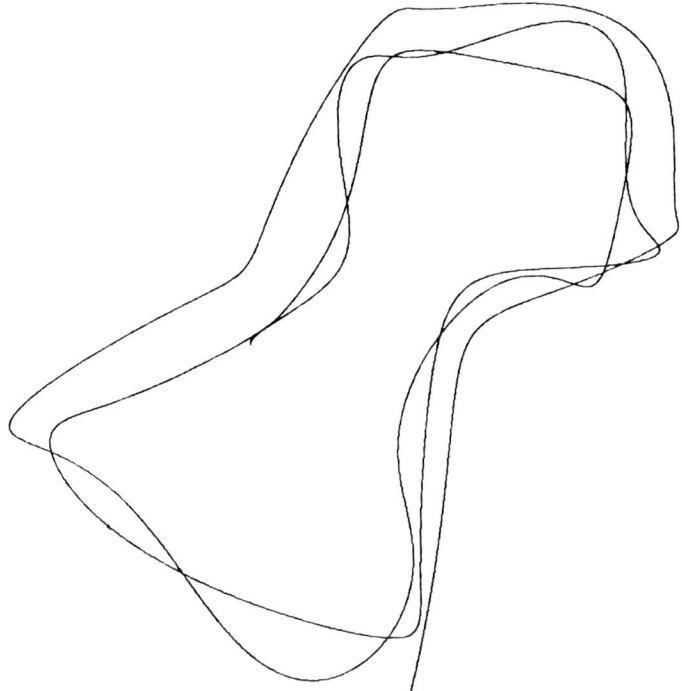

FIGURE 5. A highly filtered version of the phase portrait of FIGURE 4 to bring out the global structure.

Qualitative modeling of the structures seen in these reconstructions would be aided by estimates of the dimensions of the apparent attractor. We have computed fractal dimensions for the solar attractor using standard methods[8] and an embedding dimension of eight. The estimates converge quite well and the scaling curves from which the dimensions are obtained in the usual procedures are quite smooth. For the unfiltered data of FIGURE 2, we get a fractal dimension of the order of five. We judge this dimension to be too high to be reliably estimated with the existing data.[9] Our present work in this field is concerned with trying to assess the meaning of this result in terms of current methods for calculating dimensions. In particular, the number of circuits around what we take to be a solar attractor is too small for our purposes.[10]

Though the data we have used do not include the Maunder intermission, it seems worthwhile to try models that produce the sunspot numbers for the period when they are best known. The rough indications of our study are that it would be reasonable to try to represent the existing data with a fifth-order system. In fact, Ruzmaikin[2] has suggested that one might make do with the third-order Lorentz equations because these model a disk dynamo.[11] It is perhaps of interest that an extension of such models to five dimensions qualitatively reproduces events like the Maunder intermission.[12,13] Other possibly useful fifth-order systems have been studied in the context of dynamo models,[14,15] and such studies will help in evaluating the application of chaos theory to the solar cycle. It will be interesting to try to see whether such models have any advantage over completely stochastic models (e.g., see reference 16).

REFERENCES

1. GUDZENKO, L. I. & V. E. CHETROPRUD. 1976. *In* The Kinematics of Simple Models in the Theory of Oscillations. N. G. Basov, Ed. Proc. P. N. Lebedev Physics Institute 90. (English version by: 1978. Plenum. New York; see chapt. VIII, p. 153.)
2. RUZMAIKIN, A. A. 1981. Comments Astrophys. Space Phys. **9**: 85.
3. EDDY, J. A. 1972. Science **192**: 1189.
4. SPIEGEL, E. A. 1977. *In* Problems of Stellar Convection. E. A. Spiegel & J-P. Zahn, Eds.: 3. Lecture Notes in Physics 71. Springer Pub. New York.
5. (a) BRACEWELL, R. N. 1985. Aust. J. Phys. **38**: 1009; (b) BRACEWELL, R. N. 1953. Nature **133**: 572.
6. TAKENS, F. 1981. *In* Lecture Notes in Mathematics **898**, p. 366.
7. KAISER, J. F. & W. A. REED. 1977. Rev. Sci. Int. **11**: 1447.
8. GRASSBERGER, P. & I. PROCACCIA. 1983. Physica **9D**: 189.
9. GREENSIDE, H. S., A. WOLF, J. SWIFT & T. PIGNATURO. 1982. Phys. Rev. **25A**: 3453.
10. WOLF, A., J. B. SWIFT, H. L. SWINNEY & J. A. VASTANO. 1985. Physica **16D**: 285.
11. MALKUS, W. V. R. 1972. Trans. Am. Geophys. Union **53**: 617.
12. SPIEGEL, E. A. 1980. Ann. N.Y. Acad. Sci. **357**: 305
13. SPIEGEL, E. A. 1985. *In* Chaos in Astrophysics. R. Buchler, J. Perdang & E. A. Spiegel, Eds.: 91. Reidel. Dordrecht.
14. CHILDRESS, S. & Y. FAUTRELL. 1983. Geophys. Astrophys. Fluid Dyn. **22**: 235.
15. WEISS, N. O., F. CATTANEO & C. A. JONES. 1984. Geophys. Astrophys. Fluid Dyn.
16. BARNES, J. A., H. H. SARGENT III & P. V. TRYON. 1976. *In* The Ancient Sun. R. O. Pepin, J. A. Eddy & R. B. Merrill, Eds. Pergamon. Elmsford, New York.

Strange Accumulators[a]

L. A. SMITH[b] AND E. A. SPIEGEL[c]

[b]*Department of Physics*
Columbia University
New York, New York 10027
and
NASA Goddard Space Flight Center
Institute for Space Studies
New York, New York 10025

[c]*Department of Astronomy*
Columbia University
New York, New York 10027

In astronomy, we see many irregular structures manifested in the form of strong inhomogeneities of physical variables like density, temperature, or magnetic field. The appearance of strong concentrations of these variables, in the presence of what frequently must be highly turbulent conditions, poses an interesting question whose answer probably lies in the nature of turbulence itself. Indeed, G. I. Taylor, seventy years ago, wrote that turbulence is the strong concentration of vorticity. How does this work?

Fluid dynamicists often refer to a mechanism that they call Batchelor-Prandtl expulsion to explain inhomogeneities; the nature of this process has recently been clarified by Rhines and Young.[1] Solar physicists know of the process as it applies to the development of magnetic inhomogeneities on the solar surface.[2] However, no unanimity seems to exist among solar physicists on the explanation of the fine-scale structures on the sun.[3] There even seems to be disagreement about the proper description of the topology of field lines.

Dynamical systems theory shows us a simple way to look at field lines that may teach us about their structures. Though the approach does not explain how such fields arise in hydromagnetics (MHD), it may nevertheless be valuable in our thinking about the possibilities that confront us when we look at a complicated situation like that of the solar surface.

Consider a vector field $B(x, t)$ whose structure we want to look at. This may be a magnetic or vorticity field whose field structure may be too complicated to pull out of numerical solutions of the MHD equations (for example). Thus, this is an example of how the theoretical developments of chaos theory may help in astrophysics by adding to the ways we have of thinking about problems, even when the data may not be good enough to confirm or deny the presence of chaos.

---

[a]We acknowledge support from NSF PHY 80-2371 and NASA Cooperative Agreement No. NCC 5-29.

Let X be a function of an independent variable s such that

$$\frac{dX}{ds} = B(X(s), t) \qquad (1)$$

for fixed t. If B should happen to be a velocity field, we would identify s and t and integrate equation 1 to get the particle paths (as has been done for fluid flows in two dimensions).[4,5] However, if B is some other field, we can look at a snapshot of it by regarding t as a parameter that we hold fixed during a run in which we study equation 1 as a dynamical system with s as the "time". The trajectory in X-space will give us a picture of the field. We can even go farther and cut the X-space with a plane and simply plot the points where the trajectory pierces it. By studying typical maps of the plane into itself, we can get an idea of the possibilities that lie in store with real fields, B.

Perhaps the most typical, typical map is the standard map:[6]

$$x_{i+1} = x_i - k \sin 2\pi y_{i+1}, \quad y_{i+1} = x_i + y_i. \qquad \text{mod } 1 \quad (2)$$

This is an area-preserving map that seems a qualitatively appropriate choice when (the solenoidal) B does not vary much in the direction normal to the surface of section. For given k (not too large) and for suitable initial coordinates, this map gives rise to regular behavior in the form of periodic orbits lying on tori. The cross sections of these tori in the surface of section form island chains that are like the cat's-eyes of the fluid dynamicist. Between the islands are saddle points, in the neighborhood of which there are smaller islands, and so on.

A trajectory that gets in amongst the small islands will go for many twists before reemerging into the chaotic sea. This sojourn is responsible for long-range correlations of the successive points on the orbits of equation 2.[7,8] Meiss and Ott[9] have modeled the tendency of an orbit to linger in the reefs, while MacKay, Meiss, and Percival,[10] among others, have studied the mechanism of escape from them.

Our own naive point of view about the long times spent by particles in the reefs of the map is to liken these regions to the attractors of dissipative theory. One of the simplest examples of a dynamical system with an attractor is the Landau equation for $A(t)$:

$$\frac{dA}{dt} = A - A^3. \qquad (3)$$

This system has attractors at $A = \pm 1$ and a repeller at $A = 0$. If we differentiate this system once and substitute, we get

$$\frac{d^2 A}{dt^2} = -\frac{\partial V}{\partial A}, \qquad (4)$$

where

$$V = -\tfrac{1}{2} A^2 (A^2 - 1)^2. \qquad (5)$$

The original system (equation 3) is contained in equations 4 and 5 if we impose suitable initial conditions.

Where the original Landau equation had either attractors or repellers, the new system (which is Hamiltonian) has saddle points. In the neighborhood of these points, the motion is very slow. We call sets of these points accumulators because a representative point moving in this system will tend to spend more time in the neighborhood of these points. Hence, the points in an orbit will tend to accumulate there. In systems with stochasticity and island chains, we can have quite a complex of such points; together, these make an accumulator with very fine texture. The presence of such complicated accumulators, we suggest, gives rise to strong inhomogeneities in field structures.

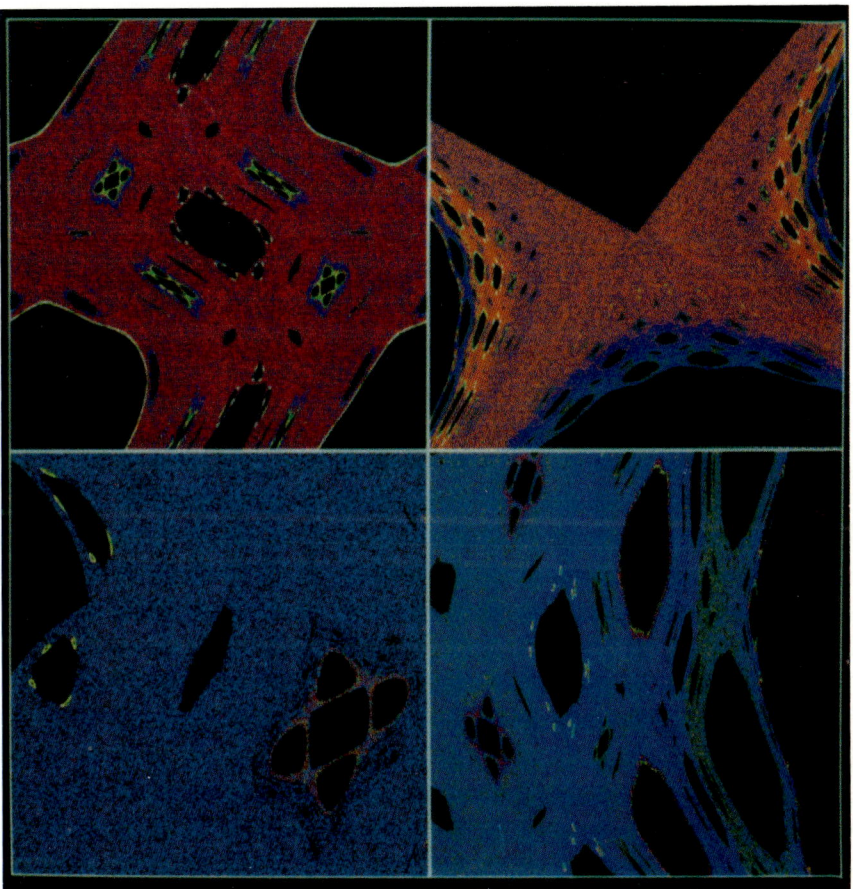

FIGURE 1. The panels on the left show the visitation histogram for the numerically one-to-one standard map ($k = 1.25$). The lower panel is a blowup. On the right, simulations with the Hénon[13] conservative map ($\cos \alpha = 0.240$) are shown. The number of visitations, as indicated by each color, varies slightly from panel to panel for photographic reasons. In the upper left panel, black means no visitation, brown is 1 to 4, blue is 5 to 9, red is 10 to 19, green is 20 to 29, and yellow is 30 to 50. White partitions were visited between 51 and 316 times. All histograms are calculated on a 2^9 by 2^9 grid.

In the left panels of the accompanying figure (FIGURE 1), we show the result of a long numerical integration based on the map of equation 2. The calculation for the standard map was done on a finite ($2^{20} \times 2^{20}$) mesh that preserves the 1–1 character of the map of equation 2, as in the work of Rannou[11] (see Miller and Prendergast[12] for the use of "noise-free" methods in dynamical astronomy). The result of this method (with a large, but finite number of grid points) is that every orbit calculated is periodic. Indeed, any deterministic simulation on a digital computer tends, in finite time, to a periodic orbit. If we are using this result to visualize field lines, this means that we are computing only closed field lines. In this case with an explicit grid, phase volumes are exactly conserved.

FIGURE 1 shows coarse-grained histograms of the number of visits to each grid square obtained in two different simulations. The panels on the left show histograms from a periodic 2,764,949-iteration orbit for the numerically one-to-one standard map with $k = 1.25$ (the lower panel being a blowup on a finer grid). The panels on the right similarly illustrate a double-precision calculation for the Hénon area-preserving, quadratic map (the field chosen is after figure 14 of Hénon[13]). In the Hénon case, we have followed the mapping for 2^{27} iterations. The calculations for the Hénon case, unlike those for the standard map, are not one-to-one numerically; hence, the structure shown in the right panels may be transient. The range in density shown is a factor of several hundred. Already, from these first results, we can see why the discussants at the solar physics meeting were in some doubt about what to call a flux tube in the solar surface. The notion of flux tube will often be useful only locally, but then the tube may splay out into the surrounding stochastic regions.

The deficiency in this type of study is that it is a little too generic. We want to be able to derive actual maps for real situations. This can be done when the vector field being studied is a real velocity field. Indeed, Chaikin et al.[5] have set up real flows corresponding closely to their maps and have found excellent agreement for the motion of advected particles. The flow in that case involved time-dependent open streamlines. We may also get open streamlines by studying the motion of particles drifting through a convecting fluid.[14,15] These studies suggest formation of strong inhomogeneities in the particle densities, but the results are, as yet, still limited.

In summation, we can say that this note is a conference publication *par excellence*. We are suggesting a way for modeling processes that seem too complex for detailed simulation without yet offering any quantitative predictions for astrophysics. Nevertheless, our qualitative results from the numerical experiments on long-period orbits seem to have surprised many of those who have seen them. Thus, it seems worth pointing out our conclusion that complicated accumulators are at the origin of strongly inhomogeneous concentrations of advected fields in flows. These fields need not be passive. Though the fields are dynamic and feed back on the flow, we may still expect the structure to be given by maps, even if we do not know them explicitly. This is ultimately a useful way to think about inhomogeneous fields.

ACKNOWLEDGMENTS

We are very greatful to B. Bayly, J. Meiss, R. MacKay, and K. Prendergast for illuminating discussions. We look forward to further illumination as we continue to pursue particles around the plane.

REFERENCES

1. RHINES, P. B. & W. R. YOUNG. 1983. J. Fluid Mech. **133**: 133–145.
2. PARKER, E. N. 1979. Cosmical Magnetic Fields. Chap. 16. Oxford Univ. Press (Clarendon). London/New York.
3. SCHMIDT, H. U., Ed. 1985. Theoretical problems in high resolution solar physics. Max-Planck-Inst. Phys. Astrophys. report no. MPA 212.
4. AREF, H. 1984. J. Fluid Mech. **143**: 1–21.
5. CHAIKEN, J., R. CHEVRAY, M. TABOR & Q. M. TAN. 1986. Proc. R. Soc. London **A405**: 165–174.
6. LICHTENBERG, A. J. & M. A. LIEBERMAN. 1983. Regular and Stochastic Motion, p. 218. Springer-Verlag. New York/Berlin.
7. KARNEY, C. F. F. 1983. Physica **8D**: 360–380.
8. CHIRIKOV, B. V. & D. L. SHEPELYANSKY. 1984. Physica **13D**: 395–400.
9. MEISS, J. D. & E. OTT. 1985. Phys. Rev. Lett. **55**: 2741–2744.
10. MACKAY, R. S., J. D. MEISS & I. C. PERCIVAL. 1984. Physica **13D**: 55–81.
11. RANNOU, F. 1974. Astron. Astrophys. **31**: 289–301.
12. MILLER, R. H. & P. H. PRENDERGAST. 1968. Astrophys. J. **151**: 699.
13. HÉNON, M. 1969. Q. Appl. Math. **27** (no. 3): 291–312.
14. SMITH, L. A. & E. A. SPIEGEL. 1985. *In* Macroscopic Modelling of Turbulent Flows. U. Frisch, J. B. Keller, G. Papanicolaou & O. Pironneau, Eds.: 306–319. Lecture Notes in Physics **230**. Springer-Verlag. New York/Berlin.
15. PASMANTER, R. A. *In* Proc. Int. Symp. Phys. Proc. Estuaries. W. van Leussen, Ed. Springer-Verlag. New York/Berlin. In press.

On the Origin of Large-Scale Cosmological Structure

J. N. FRY

Department of Physics
University of Florida
Gainesville, Florida 32611

INTRODUCTION

Although every introductory cosmology course teaches that the universe is homogeneous and isotropic, luckily this is not strictly the case, except in a statistical sense or on very large scales. Within the visible portion of the universe, matter is organized into a hierarchy of structures—on astronomical scales extending locally from stars and globular clusters of the order of millions of stars, through galaxies and small groups, containing up to trillions of stars, to clusters and superclusters containing thousands of galaxies. It seems clear that there ought to be a wealth of information in all this structure of the universe, if only we knew how to read it. In the universe at large, tracers of structure are the "island universes", or galaxies—the fossilized dinosaur footprints of Large-Scale Structure. It is from the distribution of these that we hope to learn about the nature of the universe in the distant past and the physics that shaped it. However, on examination, it amounts to an item of faith that the present distribution of matter (as inferred from the distribution of light or galaxies) contains any discernible information about the past. First, this involves the assumption that what we can see has something to do with underlying reality, that is, that the visible galaxies are fair tracers of matter in general rather than rare anomalies; second, it involves a belief that we know what the physical forces are that shaped the reality, so we can understand and interpret what we see; and third, it involves a trust that the present condition of the universe remembers the past, so we can indeed infer backwards. Finally, from this, we would like to learn about physics at, say, 10^{19} GeV. Pierre Ramond has likened this procedure to examining the dinosaur footprints and concluding, "Aha! They had blue eyes!"

The third of these questions can be tested, at least within the framework of specific models. As for our understanding of the physics involved, our laboratory experience should have taught us that whenever we reach new regimes (of size, energy, velocity, temperature, or density), we discover new physics (cf. relativity or quantum mechanics, or, now, quantum superstrings in a 10-dimensional space-time?). Nevertheless, from ignorance, it will be assumed in the following that the only significant force acting on large scales today is gravity, as described by Newton under appropriate conditions or as modified by Einstein when $v \approx c$ or $r \approx R$ (the curvature of the universe); in addition, it will be assumed that when other forces are important, they are of the form observed or extrapolated from the laboratory. Whether galaxies trace the distribution of mass is an open question, but there does not appear to be too much room for large discrepancies.[1-3]

It should be emphasized that we do not know at this point with any certainty what is the ultimate origin of cosmological structure. However, in recent years, at least a common framework for discussing the cosmological problem in physical terms has evolved. There is a collection of assumptions that make up a more or less standard model, wherein a broad spectrum of quantum fluctuations from an early epoch, modulated by physical effects that depend on the nature of the dominant component of the mass of the universe, provide the seeds that are amplified by gravitational attraction into the structures that we see today. This at least allows some statement on what this origin is not. Although all of the individual choices involved are relatively plausible, there are many steps along the way, and the resulting construct should by no means be taken to be the only possible version of the truth. In the following, I summarize the more commonly held beliefs and outline what has come to be the standard model. Of necessity, this will more or less be a catalog that outlines main points, but with most details left to the references (which also contain some impressive visual representations of the results of numerical simulations). To this end, the next section below introduces the global background within which we live—the standard Friedmann-Robertson-Walker cosmologies; the section after that discusses the inhomogeneous structure itself; and the final section contains a status evaluation and mentions in passing some alternatives to the standard model presently under study.

BACKGROUND

Our understanding of the theory of global cosmology has been more or less unchanged now for decades. Despite local variations, it is true that in a statistical sense, or averaged over large enough volumes, the universe is homogeneous and isotropic. Such a state is well described by the Robertson-Walker line element (see, e.g., reference 4),

$$ds^2 = -c^2 dt^2 + a^2(t)\,[dx^2/(1 - x^2/R^2) + x^2 d\Omega^2], \tag{1}$$

with the behavior of the universal expansion factor $a(t)$ found from Einstein's equations, $G_{\mu\nu} = 8\pi G T_{\mu\nu}$, to satisfy Friedmann's equation,

$$\left(\frac{1}{a}\frac{da}{dt}\right)^2 = \frac{8}{3}\pi G \rho - \frac{c^2}{R^2 a^2}. \tag{2}$$

A cosmological constant can be included as a contribution to the total mass density, with $\Lambda = 8\pi G \rho_\Lambda$. The curvature of space, R^{-2}, appears in the place of the total energy in a Newtonian gravitational problem: for positive curvature, the energy is negative, and the universe reaches a maximum diameter, turns around, and recollapses; however, a negative curvature universe expands without bound. From the geodesic equation in this metric, it can be shown that all momenta are damped by the expansion; in particular, photon frequencies redshift as $\omega = \omega_0 a_0/a(t)$. Objects at rest in comoving coordinates move apart with expansion as $r = a(t)x$, and thus with a velocity proportional to distance, $v = Hr$, where H is the Hubble constant, $H = \dot{a}/a$. Matter typically obeys an equation of state of the form $p = \nu\rho c^2$ for a particular component, with $\nu = \frac{1}{3}$ for radiation, $\nu = 0$ for nonrelativistic matter ("dust"), etc. The most extreme forms of matter consistent with causality and positive energy have $\nu = \pm 1$. In Einstein's theory, we also have energy conservation, in the form

$$\frac{d}{dt}(\rho a^3) + \frac{p}{c^2}\frac{d}{dt}(a^3) = 0. \qquad (3)$$

Thus, the energy density of a component with an equation of state $p = \nu\rho c^2$ scales with expansion as $\rho \sim a^{-(3+3\nu)}$, and a universe dominated by one such component expands as a power law with time, $a(t) \sim t^{2/(3+3\nu)}$. At any time, there is a causal horizon, which is the maximum distance that a null (light) signal can have traveled since $t = 0$. For a universe dominated by a species with a given ν, this is

$$r_H = a(t)\int_0^t c\,dt'/a(t') = \frac{(3+3\nu)}{(1+3\nu)}ct, \qquad (4)$$

except for $\nu = -1$, for which r_H is not zero, but for which the region of space that a given observer can communicate with is shrinking as time passes. The presence of such an horizon may cause concern when it is recognized that the 3 K background radiation is isotropic (to a part in 10^4 or better) from regions in space that could not have talked to each other before the time of last scattering ($z \approx 1500$) in the standard Hot Big Bang, which is radiation dominated ($\nu = 1/3$) from $t = 0$ until recent time. This is a first indication that there may be some modification necessary to the simplest Hot Big Bang model, which is dominated by "normal" radiation since $t = 0$ (see reference 5).

There are other difficulties in the simplest Hot Big Bang. For a given expansion rate, $H = a^{-1}da/dt$, we can define a critical density for which the curvature would vanish—this being $\rho_c = 3H^2/8\pi G$. This creates an uncomfortable coincidence, as follows. The various matter and curvature terms on the right-hand side of equation 2 scale differently with expansion. If we define Ω to be the ratio of the average density to the critical density, then it follows that

$$\Omega^{-1} - 1 = (\Omega_0^{-1} - 1)\left(\frac{\bar{\rho}_0 a_0^2}{\bar{\rho} a^2}\right), \qquad (5)$$

where the subscript zero indicates the present. The right-hand side goes as $(1 + z)^{-1}$ in a matter-dominated universe or as $(1 + z)^{-2}$ when radiation dominated. Thus, we see that as the universe expands and matter dilutes, Ω evolves away from the critical value $\Omega = 1$. Observations[6,7] typically produce $\Omega_0 \approx 0.2$–0.3. We must then live in a special time to be in a state with $\Omega \approx 1$ approximately, but not precisely. A different way to state the degree of coincidence is to notice that when the temperature of the universe was the Planck mass, $T \approx m_{pl} = 10^{19}$ GeV, say, the potential and kinetic energies had to be adjusted to match to within one part in 10^{56} in order that $\Omega_0 \approx 1$ now, even if Ω_0 is as small as 0.01.

One resolution of this discomfort is found in the modification of the simplest Hot Big Bang known familiarly as inflation.[5,8,9] In the usual realization of this model, it is presumed that at some early time, a significant component of ρ was due to a vacuum energy, that is, $T_{\hat{\mu}\hat{\nu}} = \rho_0\eta_{\hat{\mu}\hat{\nu}}$ ("hats" denote "physical components"), which is a contribution with $\nu = -1$. This type of component is unique in that the density ρ_0 is independent of epoch; as the universe expands, all other contributions to ρ quickly redshift away and the expansion factor approaches an exponential, $a(t) = \exp(\chi t)$, with $\chi^2 = 8/3\pi G\rho_0$. The curvature term in equation 2 redshifts away as well; if the exponential phase lasts for 60 scale times or so (about 10^{-35} s), the universe becomes as flat as we could want. Any inhomogeneities are expanded to extremely large scales as well.

To reach our own universe from this rapidly expanding state, we must exit from this phase somehow. In particular models under study, this occurs when a delayed phase transition occurs, thus breaking the Grand Unification symmetry; in this case, the vacuum energy ρ_0 is released as a very homogeneous thermal distribution of particles, with a temperature being such that afterwards $\mathcal{N}aT^4 = \rho_0$ (where $\mathcal{N}$ is the effective number of bosonic degrees of freedom). This model predicts that $\Omega_0 = 1$ today to a high degree of accuracy and explains why T can be constant in the microwave background, even over scales that were never in causal contact in the simpler model. A bonus at this stage is that the same Grand Unified models also predict that a matter-antimatter asymmetry (such as the observed excess in the universe of about 1 proton or neutron per 10^{10} photons in the microwave background) can arise spontaneously and naturally in an initially symmetric universe.[10] One beauty of inflationary models is that the predictions are universal and robust, and are not dependent on the particular details of a particular realization.

STRUCTURE

Origin

To many of us, a subject more interesting than the homogeneous background is the nature of departures from homogeneity. First, we need to know how to characterize these. In one frequently encountered model, the density field is treated as a fluctuating random variable, with our universe being one particular realization. Although we have only one sample, the universe is taken to be statistically homogeneous so that averages over sufficiently large volumes (a "fair sample of the universe") are equivalent to averages over the underlying probability distribution. Fluctuations are characterized statistically, for instance, by the irreducible N-point moments or correlation functions of density contrast $\delta\rho/\rho$.[11] The most fundamental two-point correlation function is often denoted $\langle \delta(\mathbf{x}_1)\delta(\mathbf{x}_2)\rangle = \xi(|\mathbf{x}_1 - \mathbf{x}_2|)$, and its Fourier transform, the power spectrum, $\langle |\tilde{\delta}(\mathbf{k})|^2\rangle = P(k)$. The latter is often preferred because different Fourier components are statistically independent, even in a highly correlated distribution. The observations are easy to characterize: the galaxy correlation function is a power law, $\xi(r) \approx (r/r_0)^{-\gamma}$, with $\gamma \approx 1.8$ and $r_0 \approx 5\ h^{-1}$Mpc, over the regime $10^3 \gtrsim \xi \gtrsim 1$. Higher order functions are also detected, with decreasing certainty, and are found to obey $\zeta_{123} = \langle \delta_1 \delta_2\ \delta_3 \rangle \approx Q(\xi_{12}\xi_{13} + \xi_{12}\xi_{23} + \xi_{13}\xi_{23})$, with $Q \approx 1$, and $\eta_{1234} \approx R[\xi_{12}\xi_{23}\xi_{34} +$ (sym.)] (16 terms in all, for all connected products of three two-point functions), with $R \approx 1$ as well.[12,13] This pattern is often referred to as "hierarchical".

Now, we must face the question of the origin of structure. Gravitational instability amplifies preexisting seed fluctuations, but only if they are large enough can they grow into what we see. The completely uniform state, although unstable, is in equilibrium; departures from homogeneity and isotropy break this symmetry and thus cannot arise from nowhere. "Shot noise", or particle discreteness, would give $\Delta N/N = 1/\sqrt{N} \approx$

10^{-34} on the scale of galaxies (galaxies have masses of order $10^{12}\,M_\odot$, where $M_\odot \approx 2 \times 10^{33}$ g is a solar mass), which is way too small ever to lead to nonlinear structure. "Thermal fluctuations" of an ideal gas are identical to particle discreteness (although interesting effects may occur near critical points of phase transitions). At present, the best guess is that structure arises from quantum fluctuations of (free) quantum fields at the time of inflation. Indeed, Stephen Hawking points out that the uncertainty principle forbids an absolutely homogeneous state of the universe. During inflation, there is nothing that can distinguish one time from another; the de-Sitter universe has a time-translation invariance. Thus, we expect the primordial spectrum of fluctuations to have no scale as well, that is, a power-law power spectrum $P(k) \sim k^n$ for some index n. Calculations[14] show that the resulting spectrum indeed reflects the de-Sitter time symmetry, with constant mean square $(\delta\rho/\rho)$ on the scale of the horizon when leaving the horizon in the de-Sitter phase, and therefore also $(\delta\rho/\rho)_H = \epsilon$, independent of scale, when reentering the horizon as well. This translates into such a power-law form, with spectral index $n = 1$ on large scales at late times, when the universe is matter-dominated. Although theory has not yet successfully predicted the value of ϵ (unlike the shape, the amplitude is highly model dependent[14]), the form of the spectrum is robust. It has been argued for some time[15,16] that this is the only reasonable form for the shape of primordial fluctuations because all other power-law spectra have divergent curvature fluctuations at small or large scales. From observational restrictions, we can take $\epsilon \approx 10^{-4}$–10^{-5} to be consistent with the small microwave temperature fluctuations, but still with an amplitude large enough to make galaxies. The primordial fluctuations are usually presumed to be Gaussian and are thus characterized entirely by the power spectrum alone. Any existing phase coherence would be lost in the long interval between the end of inflation and the time of decoupling, when fluctuations become unstable to gravitational amplification. Present structure is decidedly non-Gaussian, which must be accounted for in the theory.

Linear Evolution and Modulation

Because the problem of the evolution of structure involves much less symmetry, it is correspondingly more difficult to solve than that of the background global cosmology; the two main avenues of approach have been perturbation theory and numerical simulation. In the Newtonian regime (already difficult enough), departures from homogeneity $\delta(\mathbf{x}, t) = [\rho(\mathbf{x}, t) - \bar{\rho}]/\bar{\rho}$ obey a second-order time evolution equation,[11]

$$\frac{\partial^2 \delta}{\partial t^2} + \frac{2}{a}\frac{da}{dt}\frac{\partial \delta}{\partial t} + 4\pi G \bar{\rho}\delta = 4\pi G \bar{\rho}\delta^2 + \frac{1}{a^2}\nabla_i\phi \cdot \nabla_i\delta + \frac{1}{a^2}\nabla_i\nabla_j[(1 + \delta)v^i v^j + P^{ij}], \quad (6)$$

in comoving coordinates, with auxiliary quantities—the peculiar velocity (relative to the Hubble flow) $\mathbf{v}$ and peculiar (fluctuating) potential ϕ—determined by

$$\frac{\partial \delta}{\partial t} + \frac{1}{a}\nabla \cdot (1 + \delta)\mathbf{v} = 0 \quad (7)$$

and

$$\nabla^2 \phi = 4\pi G \bar{\rho} a^2 \delta. \quad (8)$$

These equations can be derived from a fluid mechanical model, where **v** is the velocity field and P is the pressure tensor; or statistically, from velocity moments of the Vlasov equation, with **v** as the mean and P^{ij} as the dispersion of the velocity distribution function. To linear order, equation 6 becomes

$$\ddot{\delta} + 2(\dot{a}/a)\dot{\delta} - (4\pi G\bar{\rho} - c_s^2 k^2)\delta = 0, \tag{9}$$

where c_s^2 is the speed of sound, $dp/d\rho$. This says that those modes with wavelengths that are too long for pressure response (propagating at the speed of sound) to halt collapse within a collapse time will then grow unstably; this is the Jeans instability, and the instability criterion is given by $k^2 < k_J^2 = 4\pi G\bar{\rho}/c_s^2$.[11] Before decoupling, the universe is radiation-dominated, and the Jeans mass (the mass within a volume of diameter $\lambda_J = 2\pi/k_J$) is of the order of the horizon; this, the smooth radiation distribution, and the Compton drag prevent structure from forming. (It is an interesting and unexplained coincidence that the time of equal matter and radiation densities, and the time of recombination, when the temperature is low enough for protons and electrons to form neutral hydrogen so that the universe becomes transparent, are roughly simultaneous.) At later times, when pressure is negligible ($c_s \to 0$), if the universe is cosmologically flat ($\Omega = 1$) and matter-dominated [$\nu = 0$, $\rho \sim a(t)^{-3}$], then $a(t) \sim t^{2/3}$ and $\delta(\mathbf{x}, t) = A(t)\delta(\mathbf{x}, t_0)$, with $A(t) \sim t^{2/3}$ as well.

The primordial inflationary spectrum is subject to modulations by physical processes in the intervening time between the end of inflation and the epoch of decoupling. The relativistic theory tells us that fluctuations do not grow "outside the horizon", so for the most important times, when the density contrast approaches and exceeds unity, the nonrelativistic theory is adequate. For small fluctuations, as expected at early times from small microwave temperature fluctuations, we can linearize equation 6 (or if necessary, its relativistic equivalent) and the main effect appears in a linear transfer function.[17] This is where the effects of particle physics enter; the nature of this modulation depends on the dominant contributor to the universal mass density. In this area, two different generic classes of modulations are particularly robust, with consequences not tied to one particular particle species. The so-called "hot" particles, typified by neutrinos with mass $m_\nu \approx 30$ eV, are so named because they are highly relativistic at early times and cannot be gravitationally bound until their characteristic temperature redshifts below their mass (for more discussion of hot models, see, e.g., reference 18). Fluctuations carried by these particles on scales that enter the horizon when their velocities are relativistic are strongly damped; on the other hand, those that enter later, when nonrelativistic, cannot damp, and are preserved and amplified. Thus, in the hot model, the spectrum becomes[17] $P(k) \sim k \exp[-(k/k_c)^{3/2}]$, with a strong feature at k_c, where k_c is the wave number of the horizon size at the time when $T = m$. This corresponds typically to the size of a rich cluster of galaxies, say $10^{15} M_\odot$. The so-called "cold" particles, or "cold dark matter" (CDM), including possibilities such as the pseudo-Goldstone axion, the lightest stable supersymmetry partner (photino or gravitino), or primordial black holes, are nonrelativistic at all times (for the many virtues of CDM models, see, e.g., reference 19). Their spectrum nevertheless experiences a kinetic modulation because those modes that enter the horizon when the universe is still radiation-dominated are inhibited from growing until the time of matter domination, and these short wavelength modes pile up at $P \approx \epsilon$ before this time. The effect of this piling up is to shift the spectrum from k^n at long wavelengths to k^{n-4} at short wavelengths;[17] the characteristic scale of the changeover is much the same as k_c in the neutrino case.

Nonlinear Evolution

After decoupling, linear perturbation theory tells us that density perturbations grow unstably and, in a high density ($\Omega_0 \approx 1$) universe, without bound. For a time, the evolution can continue to be followed in perturbation theory. One effect of the nonlinear theory is that even initially Gaussian fluctuations become decidedly non-Gaussian. In particular, the hierarchical form (see the subsection above entitled ORIGIN) is easily seen to arise in perturbation theory in any nonlinear theory as the lowest order contribution to the connected N-point function; however, the dimensionless coefficients Q, R, etc. are model dependent.[20] Even when fluctuations are quite small, this tells us for instance that large fluctuations, in the tails of the distribution, can be much more likely than would be expected in a Gaussian distribution, even when the variance is still small.[21,22] Inevitably, though, perturbation theory becomes inadequate. At present, the most successful approach to this regime is through numerical N-body integrations,[23-26] although even these have their problems. One difficulty with numerical approaches to the problem of cosmological structure is the wide range of important time and length scales once nonlinear evolution is reached; various techniques have been worked out to deal with this.[27] It should also be noted that the N-body numerical simulations and the theoretical calculations deal with very different representations of "reality"—one as a collection of discrete point masses and the other as a continuous field; neither of these corresponds to the real nature of galaxies, though, which are discrete, but extended objects, with internal structure. Nonetheless, the numerical approach has been very successful, and based on the numerical results, a number of general conclusions have been reached.

The qualitative differences in the behaviors of the development of structure in N-body simulations of hot and cold models are easy to picture and understand. In cold universes, the modulated primordial spectrum still contains significant power on small scales, and clumping proceeds from the "bottom up": small associations form first, which then agglomerate into larger ones that merge into still larger, with the process continuing through all times.[19] This produces a hierarchical distribution of structure, with something happening on some scale or another at all times. In hot universes, there are no short wavelength fluctuations present, and the distribution is correspondingly much smoother at early times. The first significant event is the formation of caustics: on surfaces where the velocity goes through zero, it does so smoothly; thus, with matter to the left moving to the right with velocity proportional to distance, and matter to the right moving to the left, it all arrives at once—a catastrophic event leading to sheets of very high density.[18] This focusing is a purely kinematical effect and is only modestly influenced by gravity. In the hot models, nothing happens before the caustics form, when essentially everything happens at once; the transition from a very smooth mass distribution to a very lumpy one is almost instantaneous. Once this happens, the visual appearance of the particle distributions in the hot models shows a distinctive cellular appearance, with a characteristic scale determined by k_c. Formation of individual galaxies must proceed by the fragmentation of these large units into smaller ones.

We would then like to evaluate the success of the various models in reproducing our universe. One's first impression is that cold-particle models just look more like our universe.[24] We would like to confirm that impression by a more objective process. Statistically, in both hot and cold N-body simulations, the two-point correlation function is observed to evolve with time.[23-25] Typically, the two-point function in the simulations is a power law, $\xi \sim r^{-\gamma}$, that steepens with time, with the power index passing through the observed value, $\gamma \approx 1.8$, at a well-defined moment, which then defines the "present" in the simulation. Once this freedom is removed, all other properties must agree with observations at this same time. Essentially every objective test to date has disfavored the hot models. Because the universe is quiescent until the caustics form, and from there on things evolve rapidly, we are led to conclude that galaxies formed very recently, perhaps at a redshift $z_f \lesssim 1$ in the hot models.[23] This is to be compared with quasar redshifts approaching $z_q \approx 4$ and redshifts of QSO metal absorption line systems $z_a \gtrsim 2$, which show that some material was processed through stars much earlier than could be true in the hot models. It also seems to be that to reconcile the ratio of the observed correlation length of galaxies [where $\xi(r) = 1$] to the damping length in the initial power spectrum, it requires an uncomfortably small value for the Hubble constant.[23] Other problems are found in the detailed nature of the matter distribution: the "present" state in the neutrino models is dominated by the very few, very rich clusters that form at the intersection of caustics and that end up containing 90% of the mass.[28] By contrast, rich clusters in our universe, as cataloged by George Abell, seem to contain about 5% of the galaxies. These superrich clusters would be strong X-ray emitters that are also not seen in the observations.[28] In our universe, Abell clusters are more strongly correlated than galaxies in general, while in the hot model simulations, galaxy and cluster correlation amplitudes are about the same; this is because essentially all galaxies are in the clusters.[29] Finally, the behavior of higher order correlations in the hot models is not consistent with the observed hierarchical pattern above.[25,26] By every one of these tests, the CDM model fares better. It is sometimes said that it has the "least wrong with it", but the agreement seems to warrant a stronger statement than that. Even so, the agreement is not perfect; in particular, velocities in the model simulations are too large compared with the data. This has led some to speculate that galaxies do not form with equal probability everywhere, but form preferentially at the highest peaks in the mass distribution; that is a game known as "biased galaxy formation".[1-3] This succeeds in eliminating the discrepancy, although at the cost of introducing an unknown threshold parameter.

To supplement the N-body simulations, there are a few nonlinear analytic calculations that can be done explicitly—one being the short-distance behavior of the two-point function in a scale-invariant universe (with $\Omega = 1$ and $P \sim k^n$ on large scales). From perturbation theory, we know that the correlation amplitude for weak clustering, $\xi \ll 1$, grows with time as $\xi \sim a(t) \sim t^{4/3}$ (equation 6). From the long wavelength spectrum $P(k) \sim k^n$, a Fourier transform thus gives a correlation function that goes as powers of both arguments, $\xi(x, t) \sim t^{4/3} x^{-(3+n)}$. In the nonlinear regime, $\xi \gg 1$, the momentum-projected BBGKY equations give an equation of continuity for the irreducible N-point function,[11]

$$\frac{\partial \kappa_N}{\partial t} + \frac{1}{a} \frac{\partial}{\partial x^i_A} v^i_A \kappa_N = 0. \qquad (10)$$

Physically stable clustering requires that the peculiar velocity just cancels the Hubble expansion,[30] that is, $v_A^i = -\dot{a}x_A^i$. With this, equation 10 becomes

$$\left[a\frac{\partial}{\partial a} - 3(N-1) - x_A^i\frac{\partial}{\partial x_A^i}\right]\kappa_N = 0. \tag{11}$$

As usual, the homogeneous operators indicate power-law solutions. The factor $N - 1$ appears because the center of mass of N particles is fixed. If the irreducible N-point function scales with size as $\kappa_N(\lambda \mathbf{x}_A) = (\lambda^{-\gamma N})\kappa_N(\mathbf{x}_A)$, then by the chain rule, the spatial derivatives are equivalent to the overall scaling $\lambda(d\kappa_N/d\lambda) = -\gamma_N\kappa_N$. Thus, in this regime, the solution is again a power law in both space and time, $\kappa_N \sim a^{3(N-1)-\gamma_N}x^{-\gamma_N}$. The scale invariance of the problem then lets us relate the small- and large-scale results. From scale invariance, we expect $\xi(x, t) = \xi(s)$, where $s = x/t^\alpha$.[30] (In perturbation theory, for instance, if this is true at linear order, it remains so for all orders.[20]) The result from linear theory, above, gives $\alpha = 4/(9 + 3n)$. This then gives $\xi \sim x^{-\gamma}$, with $\gamma = (9 + 3n)/(5 + n)$ for the small separation result for the two-point function and $\gamma_N = (N - 1)\gamma$ for the N-point function, both independent of epoch; this is a distribution following the "hierarchical" pattern as above, but for quite different reasons. This scale-invariant result is reminiscent of critical fluctuations near a phase transition; the loss of one dimension in the exponent can be traced to the fixed center of mass of a gravitationally interacting system of N particles. This describes some form of fractal distribution, with dimension $d_f = 3 - \gamma \approx 1.2$. This behavior is very different from the evolution observed in the numerical simulations; furthermore, it is not understood whether the N-body result is an artifact of the numerical procedure or whether the scale-invariant result reflects only the assumption of a nonexistent stability. From quite different assumptions, it has also been claimed that from thermodynamics applied to gravitational clustering, the index γ always evolves toward $\gamma = 2$ in the limit of "maximum entropy";[31] however, it is not clear what that means in a gravitating system (the specific heat is formally negative, indicating an instability, and the energy is unbounded from below). The N-particle distributions in this limit have been found to obey the hierarchical pattern $\kappa_N = Q_N \Sigma \xi^{N-1}$, yet again, with $N^{N-2}Q_N = (2N - 3)!!$.[32] This pattern appears more often than we can understand.[21]

DISCUSSION

It ought to be clear from the above that our understanding of large-scale structure is too weak to make any definitive statement about how things came to be. Even if we find a model with well-motivated initial conditions whose evolution leads to a universe consistent with most observations, there is no way at present to prove it is unique. Nonetheless, we can feel fairly confident in ruling out at least hot particle models, as discussed above, on a fairly large number of grounds; these include the epoch of galaxy formation, the value required for the Hubble constant, and the detailed nature of the resulting clusters, including the galaxy three-point function and the strength of cluster clustering. We also have found, from the simulations, that although on small scales, say separations less than about $5\,h^{-1}$ Mpc, the distribution is relatively relaxed and has forgotten about initial conditions, this is not so on larger scales. In particular, the scales most responsible for the visual impressions and textures in pictures of the data and in the simulations are in the regime that does remember the initial conditions. On these scales, the relative density contrasts—and thus correlation statistics—are small; however, it seems that the eye responds to ΔN and not $\Delta N/N$. This is consistent with calculations of volume occupancy probabilities in which the relevant parameter is not $\bar{\xi}$, but $\bar{N}\bar{\xi}$.[21]

The hot and cold models share the property that they both are the straightforward results of simple assumptions and known physics; in fact, both are robust, in the sense that the nature of the results is not strongly dependent on the details of particular models. In particular, a large number of different elementary particles, known or conjectured, lead to the CDM scenario if they represent the dominant mass of the universe. To date, the CDM model is the most successful in reproducing observational features. It may betray only a prejudice for simplicity, but there is something appealing about the universal scale invariant fluctuation spectrum of inflation subject to the universal CDM modulation. Of course, the standard hot and cold models by no means exhaust the possibilities under consideration, but other models are at present less firmly characterized and are thus less definite in their predictions. Among the alternative models are "cosmic string"[33,34] and "cosmic explosions".[35] String is a linear topological defect, or vortex line, in the order parameter breaking the Grand Unification symmetry in the early universe. The lowest energy state at $T \approx 0$ is the broken symmetry state, where only the photon mediates a long-range force.[10] Inside the core of the string, to a diameter $m_X^{-1} \approx (10^{15} \text{ GeV})^{-1} \approx 10^{-39}$ cm, the Grand Unification symmetry is restored, although at the cost of a large vacuum energy density $\rho_0 \approx m_X^4$; therefore, from a distance, the string appears to have a linearly distributed energy density per unit length $\mu \approx m_X^2 \approx 10^{30}$ g/cm, or in dimensionless units, $G\mu \approx (m_X/m_{pl})^2 \approx 10^{-8}$. In the early universe, the topological defects in the Higgs field appear because when the symmetry breaks, causality prevents the presence of correlations on scales larger than the horizon size. As the universe grows and the horizon expands, the field then discovers that with some probability of order unity per horizon volume (perhaps one time in twelve), it has tied itself in a knot when it tries to smooth out its kinks (to reduce its curvature energy). The ultimate seeds of stochasticity in this case, determining the initial configuration of string, are the small ("thermal"?, "quantum"?) fluctuations that pick out the local direction of the gauge symmetry breaking, which are then greatly amplified by the field dynamics. It appears that if strings "intercommute" or exchange connections when they intersect, then the distribution of masses and sizes of loops becomes, again, scale invariant; also, the exact initial distribution may be unimportant.[34] String models have only one adjustable parameter, $G\mu$, that can be fixed to give, say, the correct number density of galaxies.[33] A large success is that they then correctly predict the density of Abell clusters in the scale invariant model.[36] The main problem with all this is that it is difficult to predict the exact details of the relation between large-scale structure, presumably determined by the dynamics of loops of string, and the small-scale details, such as the inner workings of galaxies. In the explosion scenario, the early universe does not have imprinted all the fluctuations that will grow into the galaxies we see today, but just enough to form a few, perhaps supermassive objects that live briefly and spectacularly, and that then explode violently at the end of their life.[35] The blast from these initial seeds then creates another generation of objects, and perhaps another; until at the end, the universe is more or less uniformly populated with galaxies. The advantage of this picture is that we know that this is how many stars form, triggered by supernova explosions in so-called OB-associations. It appears difficult, though, to create structure on truly large scales (greater than $\sim 10\ h^{-1}$ Mpc) in this model.[37] Unfortunately, in both of these pictures, even more than in the hot and cold modulations of scale invariant inflationary fluctuations, it appears that the final appearance is only very indirectly tied to the initial conditions.

As a final caution, it should be pointed out that there are unexplained observations as well. Most data to date on the positions of galaxies are two-dimensional, that is, projected on the sphere of the sky. Much of the data has been around for decades, and recently the statistical uniformity of the older data has been questioned.[38] For both of these reasons, we would like to greatly increase our sample of modern, three-dimensional measurements. However, distances to galaxies are difficult to obtain; the simplest indicator, apparent brightness, which scales as r^{-2}, is unreliable because of the large scatter in intrinsic brightness of galaxies. Redshifts are the most useful distance indicators, through Hubble's law $v = Hr$, but they take much more telescope time than do the integrated broadband brightness measurements; furthermore, they are also contaminated by peculiar motions. Thus, we have waited for the advent of high quantum efficiency detectors and larger telescopes to try to get three-dimensional information about large samples of galaxies. Many of these reconfirm earlier results, at least out to modest distances.[6] However, on larger scales, recent three-dimensional observations unexpectedly show a "frothy" distribution, with galaxies appearing to lie on the surface of sharp-edged bubbles.[39] In addition, there seem to be unexplained large-scale streaming motions that cannot be associated with density fluctuations.[40] If they were associated, there would be detectable microwave background temperature anisotropies. While these phenomena are not completely incompatible with the standard models, they also do not fall out naturally. They seem to imply significant large-scale power in the initial fluctuation spectrum to a degree incompatible either with inflation or with the observed microwave background isotropy. Thus, it should be exciting to see how this subject continues to develop.

REFERENCES

1. BARDEEN, J., J. R. BOND, N. KAISER & A. SZALAY. 1986. Astrophys. J. **304**: 15.
2. KAISER, N. 1984. Astrophys. J. Lett. **284**: L9.
3. KAISER, N. 1986. *In* Inner Space/Outer Space. E. Kolb, M. Turner, K. Olive, D. Sekel & D. Lindley, Eds. Univ. of Chicago Press. Chicago.
4. MISNER, C. W., K. S. THORNE & J. A. WHEELER. 1973. Gravitation. Freeman. San Francisco.
5. GUTH, A. 1981. Phys. Rev. **D23**: 347.
6. DAVIS, M. & P. J. E. PEEBLES. 1983. Astrophys. J. **267**: 465.
7. DAVIS, M. & P. J. E. PEEBLES. 1983. Annu. Rev. Astron. Astrophys. **21**: 109.
8. ALBRECHT, A. & P. STEINHART. 1982. Phys. Rev. Lett. **123**: 456.
9. LINDE, A. 1982. Phys. Lett. **108B**: 389.
10. KOLB, E. W. & M. S. TURNER. 1983. Annu. Rev. Nucl. Part. Sci. **33**: 645.
11. PEEBLES, P. J. E. 1980. The Large-Scale Structure of the Universe. Princeton Univ. Press. Princeton, New Jersey.
12. GROTH, E. J. & P. J. E. PEEBLES. 1977. Astrophys. J. **123**: 456.
13. FRY, J. N. & P. J. E. PEEBLES. 1978. Astrophys. J. **123**: 456.
14. BRANDENBERGER, R. *In* Inner Space/Outer Space. E. Kolb, M. Turner, K. Olive, D. Sekel & D. Lindley, Eds. Univ. of Chicago Press. Chicago.
15. HARRISON, E. R. 1970. Phys. Rev. **D1**: 2726.
16. ZEL'DOVICH, YA. B. 1972. Mon. Not. R. Astron. Soc. **160**: 1P.

17. Bond, J. R. & A. Szalay. 1983. Astrophys. J. **274:** 443.
18. Zel'dovich, Ya. B. 1978. *In* IAU Symposium No. 79—The Large-Scale Structure of the Universe. M. S. Longair & J. Einasto, Eds. Reidel. Dordrecht.
19. Blumenthal, G., S. Faber, J. Primack & M. J. Rees. 1984. Nature **311:** 517.
20. Fry, J. N. 1983. Astrophys. J. **267:** 483.
21. Fry, J. N. 1985. Astrophys. J. **289:** 10.
22. Fry, J. N. 1986. Astrophys. J. Lett. **308:** L71.
23. Frenk, C. S., S. D. M. White & M. Davis. 1983. Astrophys. J. **271:** 417.
24. Davis, M., G. Efstathiou, C. Frenk & S. D. M. White. 1985. Astrophys. J. **292:** 371.
25. Fry, J. N. & A. L. Melott. 1985. Astrophys. J. **292:** 395.
26. Melott, A. L. & J. N. Fry. 1986. Astrophys. J. **305:** 1.
27. Efstathiou, G., M. Davis, C. S. Frenk & S. D. M. White. 1985. Astrophys. J. Suppl. Ser. **57:** 241.
28. White, S. D. M. 1986. *In* Inner Space/Outer Space. E. Kolb, M. Turner, K. Olive, D. Sekel & D. Lindley, Eds. Univ. of Chicago Press. Chicago.
29. Barnes, J., A. Dekel, G. Efstathiou & C. S. Frenk. 1985. Astrophys. J. **295:** 368.
30. Davis, M. & P. J. E. Peebles. 1977. Astrophys. J. Suppl. Ser. **35:** 425.
31. Saslaw, W. C. 1980. Astrophys. J. **235:** 299.
32. Saslaw, W. C. & A. J. S. Hamilton. 1984. Astrophys. J. **276:** 13.
33. Turok, N. 1986. *In* Proceedings of the 13th Texas Symposium on Relativistic Astrophysics. N.Y. Acad. Sci. New York. To be published.
34. Vilenkin, A. 1984. Phys. Rev. Lett. **53:** 1016.
35. Ostriker, J. & L. Cowie. 1981. Astrophys. J. Lett. **243:** L127.
36. Turok, N. & R. H. Brandenberger. 1986. Phys. Rev. **D33:** 2175.
37. Charlton, J. C. & D. N. Schramm. 1986. Astrophys. J. **310:** 26.
38. Geller, M., V. de Lapparent & M. Kurtz. 1984. Astrophys. J. Lett. **287:** L55.
39. de Lapparent, V., M. Geller & J. Huchra. 1986. Astrophys. J. Lett. **302:** L9.
40. Dressler, A., S. M. Faber, D. Burstein, R. L. Davies, D. Lynden-Bell, R. J. Terlevich & G. Wegner. 1987. Astrophys. J. Lett. **313:** L37.

Stability of an Area-Preserving Mapping

JAMES H. BARTLETT

Department of Physics and Astronomy
University of Alabama
Tuscaloosa, Alabama 35487-1921

When designing a high-energy circular particle accelerator, it is essential to ensure that the number of particles lost from the beam is minimal. This has led to the study of Hamiltonian systems with periodic coefficients and to the investigation of the stability of area-preserving mappings.

As a prototype of an area-preserving mapping, we[1,2] have studied the system

$$T: x' = x + a(y - y^3),$$
$$y' = y - a(x' - x'^3), \quad (1)$$

and have located fixed points and variational matrix traces as a function of the parameter "a" for $a \geq 1.0$. The present paper extends the range of "a" to lower values and contains a discussion of how the stable region will change in extent as "a" changes.

CALCULATION OF TRACES

The variational equations for equation 1 are:

$$\Delta x' = \Delta x + a(1 - 3y^2)\Delta y,$$
$$\Delta y' = \Delta y - a(1 - 3x'^2)\Delta x'. \quad (2)$$

For a fixed point under n mappings (T^n), let

$$\Delta x' = A \Delta x + B \Delta y,$$
$$\Delta y' = C \Delta x + D \Delta y. \quad (3)$$

If the mapping is repeated,

$$\Delta x'' = A \Delta x' + B \Delta y' = (A^2 + BC)\Delta x + B(A + D)\Delta y,$$
$$\Delta y'' = C \Delta x' + D \Delta y' = C(A + D)\Delta x + (BC + D^2)\Delta y. \quad (4)$$

To find the trace $\mathrm{tr} \equiv A + D$ of equation 3, one can iterate equation 2 n times. Alternatively, let $\Delta y = 0$ and we have $\Delta y' = C \Delta x$ and $\Delta y'' = C(A + D)\Delta x$, from which

$$A + D = \Delta y''/\Delta y'. \quad (5)$$

The use of equation 5 is often very convenient where extreme accuracy is not required and where one knows the Δy's to seven or eight places. (Whereas our older published calculations and most of those reflected in TABLE 1 of this report were performed to double precision with UNIVAC 1100/60, the recent work has been done to quadruple precision with the IBM 3203.)

TABLE 1. Traces for $a = 1.0$[a]

n/k	x_f	Trace	n/k	x_f	Trace	n/k	x_f	Trace
19/3	0.22242765	2.	34/3	0.750618	1.9528	14/1	0.84684089	−1.5826	78/5	0.888704	0.06805
13/2	0.27110084	2.	23/2	0.758184	2.	85/6	0.86848732	2372.	47/3	0.8901675	2.0140
20/3	0.31147955	2.	35/3	0.76545446	2.	71/5	0.86848802	216.4	63/4	0.891974	2.0090
7/1	0.37754935	2.	12/1	0.77899758	2.1234	57/4	0.86849572	21.47	79/5	0.89314949	2.0101
22/3	0.43130555	2.	37/3	0.79145928	2.	43/3	0.86857989	3.8435	95/6	0.89392611	2.0181
15/2	0.45497895	2.	25/2	0.797372	2.	72/5	0.86878898	4.8272	16/1	0.89581337	4.6242
23/3	0.47695886	2.	38/3	0.802975	2.	29/2	0.86944327	2.2424	97/6	0.89754174	2.2976
8/1	0.51671626	2.0012	13/1	0.81323372	2.1881	73/5	0.870494	2.0043	81/5	0.89835007	2.1514
25/3	0.55193798	2.	79/6	0.81774988	2.0056	44/3	0.87138692	2.5880	65/4	0.89939265	2.1195
17/2	0.56815066	2.	66/5	0.818584	2.	59/4	0.872664	2.0031	49/3	0.90125414	2.1626
26/3	0.58355709	1.9995	53/4	0.81978599	2.0586	74/5	0.873462	4.6358	82/5	0.90246360	15.385
9/1	0.61228814	2.	40/3	0.821630	2.0122	89/6	0.87398984	2.0002	33/2	0.90420956	2.4065
28/3	0.638851	2.	67/5	0.82292819	2.	15/1	0.87702298	2.1705	545/33	0.90457536	2.00002
19/2	0.65167754	2.	27/2	0.824491	2.0039	91/6	0.88025777	2.00003	347/21	0.90457536	2.0008
29/3	0.66465313	2.	68/5	0.825481	2.0348	76/5	0.88094636	2.0059	149/9	0.90502647	2.1523
10/1	0.68140488	3.2369	41/3	0.825805	2.0246	61/4	0.882037	2.0034	116/7	0.905217	5.0882
31/3	0.696180	2.	55/4	0.82593798	2.2525	46/3	0.88339425	7.9889	199/12	0.9053287	2.9737
21/2	0.706945	2.	69/5	0.82595018	4.7701	77/5	0.884354	2.0035	83/5	0.90542117	4.5268
32/3	0.716723	2.	83/6	0.82595129	32.6	31/2	0.886567	2.0399	50/3	0.9080061	−1.5383
11/1	0.73449616	2.0004									

[a]When the trace is entered as "2.", the fixed point is hyperbolic and the trace is close to and greater than 2.0. (The exact value has not been shown in order to save space.)

BEHAVIOR OF TRACES

If, for a fixed point of order n, the trace tr (T^n) is less than 2.0, then the point is elliptic and points nearby are mapped around the fixed point in approximate ellipses.

If, on the other hand, the trace tr (T^n) is greater than 2.0, the fixed point is hyperbolic. Points in the neighborhood are mapped along approximate hyperbolae, asymptotic to eigencurves from the fixed point. Motion is outward along one pair of eigencurves and inward along the other pair.

For a given value of n, the trace goes asymptotically to 2.0 as the parameter a is decreased. Elliptic points remain elliptic and hyperbolic points remain hyperbolic.

Fixed points lie along the lines $y = 0$, $x = 0$, $x = y$, $x = -y$, or are maps of such points.

MECHANISMS OF ESCAPE

Let us suppose that we have located a hyperbolic fixed point P of order n on the $+x$-axis and proceed counterclockwise toward the next fixed point Q, also of order n. At P, one eigencurve EP comes in from the left and another eigencurve PF departs to the right of this. At Q, one eigencurve QA departs (toward P) on the left and another eigencurve HQ comes in on the right. These four curves enclose an elliptic point between P and Q and constitute roughly an "island" or a "tangle". For each value of n, the origin is surrounded by a chain of such islands, much like the links of a sausage. (A similar picture holds if the fixed point on the $+x$-axis is elliptic and the hyperbolic points are elsewhere.)

The outer curve PF will intersect the outer curve HQ at a homoclinic point J. When PF has proceeded past J, it oscillates about HQ. The areas outside of HQ and inside PF will be compressed toward the outward eigencurve from Q toward the next fixed point R. Eventually, the origin will be encircled and a long and thin tongue will cross the x-axis outside of P. Thus, whenever the trace is greater than 2.0, there will be constant spiraling outward motion from the corresponding chain of islands. The motion will persist until an inner eigencurve of a neighboring chain is encountered (or there may be a direct intersection without spiraling). Motion then occurs along this inner eigencurve until the tangle (island) is entered; it continues toward a hyperbolic point, eventually describing a hyperbola and following the outward eigencurve from this hyperbolic point. Such processes, repeated, enable escape to occur. This could be pictured as a "bucket brigade".

A practical way of finding a region around the origin that is almost stable suggests itself. For each fraction n/k (where n is the order of the fixed point and k is the number of times that the origin is encircled), we can construct the corresponding chain of islands and can estimate intersection angles at the homoclinic points. If such an intersection angle is very small, as it will be when the trace of the variational matrix deviates from 2.0 by a very small amount, the island will be very thin, and the area available for mapping outward (or inward) will be very small relative to the area between the chain of islands and the origin. Outward motion will be extremely slow. When the trace of the variational matrix is close to 2.0, the region interior to the chain of islands is almost stable.

In this report, we have extended previous work in two directions. First, for a fixed value of the parameter "a", namely, $a = 1.0$, we have more details about fixed points on the x-axis (and have found some with values of the trace close to 2.0) for values of x somewhat below our previous value of $x = 0.905468199$, from which escape to infinity occurs in about 40,000 mappings. Secondly, we have varied the parameter "a", and for each value of "a", we have found the apparently outermost fixed point with trace close to 2.0, thus delimiting the "almost stable" region. Noting that the trace approaches 2.0 asymptotically as "a" decreases, we expect that further examination of the behavior as "a" increases will provide detailed insight into how the transition to chaos (i.e., the breakup of "invariant" curves with trace equal to 2.0) occurs.

For $a = 1.0$, TABLE 1 shows the general behavior of the traces for representative

TABLE 2. Fixed Points on the x-Axis That Mark the Apparent Outer Edge of the "Almost Stable" Region[a]

a	n/k	x_f	Trace
0.3	180/1	0.99989126	2.0000016
0.4	100/1	0.99867721	2.00047
0.5	65/1	0.99489983	2.000033
0.6	45/1	0.98600357	2.0015
0.7	29/1	0.95358657	2.00035
0.8	21/1	0.91127828	2.00077
0.9	77/1	0.91424557	2.00079
1.0	545/33	0.90442347	2.0000198
1.1	35/3	0.80082635	2.000158
1.2	17/2	0.68560821	2.000024
1.3	33/4	0.71048461	2.000079
1.40	109/1	0.61885203	2.000000075
1.42	109/1	0.63210404	2.000308
1.43	17/1	0.45581072	2.000253
1.50	27/1	0.45969238	2.000000006
1.6	27/1	0.47792645	2.000575
1.7	64/1	0.45677176	2.00000526
1.8	31/1	0.41307839	2.0000015
1.9	89/1	0.42843742	2.000126
2.0	61/1	0.35867772	2.000130

[a]These are preliminary results, which may be improved somewhat on closer examination.

fixed points (small values of n and k) on the x-axis from near the origin out to the region of instability. The coordinates of a fixed point are $x = x_f$, $y = 0$. The fixed points shown here have traces very close to 2.0 as x_f becomes small. Below $x = 0.82$, there are fairly large intervals where the traces are close to 2.0. (These are interrupted by points with appreciable deviations from 2.0, namely, for $n/k = 10/1$, $34/3$, $12/1$, and $38/3$.) As x increases above $x = 0.82$, such intervals become narrower, but where they exist, they constitute effective barriers against motion across them. Between the fixed points for $n/k = 33/2$ and $n/k = 116/7$, there are several points with trace close to 2.0, namely, for the series with $n/k = 347/21$, $413/25$, etc.

Having determined the apparent outermost value of x_f for which the trace is close to 2.0, it is then simple to vary the value of "a" and to scan the x-axis to see where points go to infinity for a small number (say, 500) of mappings. Additional scanning for lesser values of x readily reveals fixed points with traces close to 2.0. TABLE 2 lists such fixed points for values of "a" from 0.3 to 2.0, with values of n/k, x_f, and the trace. FIGURE 1 is a graph of this x_f versus "a", and it shows roughly the maximum extent of the stability region. At $a = 0.30$, x_f is near 1.0. This value decreases rather slowly to about 0.904 at $a = 1.0$ and then decreases more rapidly to 0.632 at $a = 1.42$. At this stage, a rather violent change occurs, so the outer bound at $a = 1.43$ appears to be near $x_f = 0.456$. There is a rise to $x_f = 0.477$ at $a = 1.60$, followed by a decrease to $x_f = 0.359$ at $a = 2.00$. (Note that there is an outer barrier, even though the origin is unstable, at $a = 2.00$.) We thus have a rough picture of how the stability region changes with "a". Further work may modify the details somewhat in the interval $1.43 \leq x \leq 2.00$ and is definitely necessary in order to produce an understanding of occurrences in the interval $1.42 \leq x \leq 1.43$.

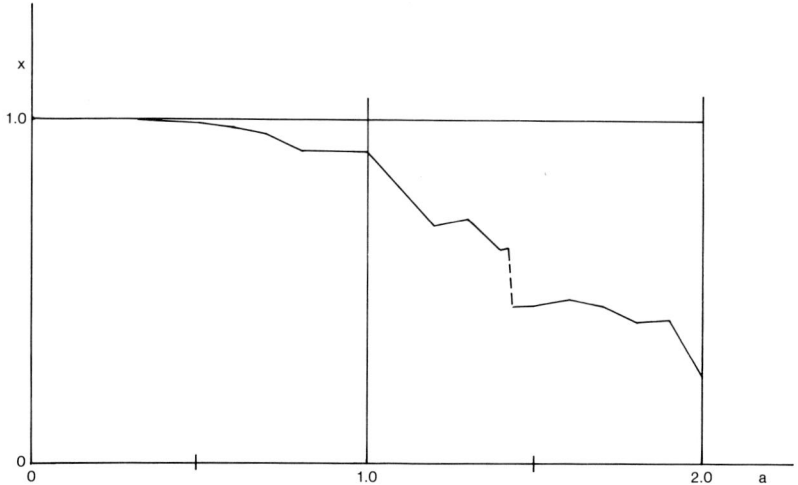

FIGURE 1. x versus a, where x is the intersection of the outermost boundary of the almost stable region with the x-axis, and "a" is the parameter of equation 1.

ACKNOWLEDGMENTS

We wish to express our thanks to G. Contopoulos for his invitation to participate in the University of Florida Workshop, and to G. Schmidt for a seminal discussion on this occasion.

REFERENCES

1. BARTLETT, J. H. 1978. Celest. Mech. **17:** 3–36. See also Errata.
2. BARTLETT, J. H. 1982. Celest. Mech. **28:** 295–317.

Resonances Fill Stochastic Phase-Space[a]

J. D. MEISS

Institute for Fusion Studies
University of Texas
Austin, Texas 78712

INTRODUCTION

This is a presentation of recent results of collaborative research with R. S. MacKay and I. C. Percival.[1,2] This research initiates a program to determine the transport properties of stochastic orbits in nonintegrable Hamiltonian systems. While most of the motion for an integrable Hamiltonian system is quasi-periodic and confined to invariant tori, a typical Hamiltonian is nonintegrable. Furthermore, although there may exist some invariant tori, many are destroyed and replaced by invariant Cantor sets, called "cantori".[1] Often, much of the remaining phase-space is filled with orbits that appear chaotic, or stochastic, and these orbits can leak through the cantori.

In our first paper (reference 1), we introduced the notion of flux for the case of area-preserving maps and showed how the flux through cantori could be calculated. Transport in chaotic regions is impeded by partial barriers that are formed on the framework of the cantori. Flux across these partial barriers takes place through "turnstiles". Choosing a discrete set of partial barriers (corresponding to the "most important" cantori) partitions the phase-space into regions between which flux occurs, with the amount being calculable as the difference in action of certain orbits. This allows one to estimate transport rates.

I will describe here the results of the second paper (reference 2) where we provide a less arbitrary partition of the phase-space. The corresponding partial barriers are formed from pieces of stable and unstable manifolds of hyperbolic periodic points, and the regions that they bound are called "resonances".

Consider an integrable system, for example, the simple pendulum. The unstable or hyperbolic orbit corresponds to the pendulum at rest with the bob on top ($\theta = \pi$). This orbit has an unstable manifold, which corresponds to that orbit beginning at rest infinitesimally close to $\theta = \pi$. The orbit is identical to the stable manifold of the hyperbolic fixed point, which corresponds to the orbit beginning at $\theta = 0$ with just enough energy to approach the vertical position as $t \to \infty$. Together, the stable and unstable manifolds form the separatrix. The resonance is defined as the region of phase-space interior to the separatrix. It contains an orbit of the same period as the hyperbolic orbit, which in this case is the stable equilibrium point. This point is surrounded by "trapped" or "vibrational" invariant curves, which make an island in phase-space. In general, when the hyperbolic orbit is periodic with period n, the resonance is made up of a chain of n such "islands".

---

[a]This international collaboration was sponsored by NATO Grant No. RG 85/0461. J. D. Meiss also received support from US DOE Grant No. DE-FG05-80ET-53088.

Similar resonances occur in profusion when integrable systems are perturbed and become nonintegrable. However, the resonances do not appear to be well defined because narrow bands of apparently chaotic motion replace the well-defined separatrices. When the perturbation amplitude is increased, the sizes of two such resonances (as calculated by perturbation theory) can become large enough so that they would overlap. This suggests that all the invariant tori between the resonances are destroyed and there is "global chaos". This is the basis of the Chirikov resonance overlap method.[3]

In the next section, we recall a precise construction for resonances of nonintegrable area-preserving maps.[4,5] These resonances are bounded by "partial separatrices" that are formed from stable and unstable manifolds of hyperbolic periodic points. They have turnstiles just like the partial barriers formed from cantori. The flux in and out of resonances takes place only through these turnstiles. The area entering or leaving a resonance from above or below is related to the difference of action between pairs of homoclinic orbits.[1]

In reference 2, formulae for the resonance areas and turnstiles are obtained in terms of the action of the homoclinic orbits. A resonance has a well-defined area regardless of whether its central periodic orbit is elliptic or hyperbolic with reflection; however, not all this area may be accessible to a chaotic region. The resonance will be called elliptic or hyperbolic, accordingly.

In the third section, the life history of a resonance with variation of a parameter representing strength of nonlinearity is discussed. Remarkably, the resonance area reaches a maximum just after the transition from the elliptic to hyperbolic case. Above this parameter value, its area decreases rapidly. The sum of all of the resonance areas in a region with no invariant circles is shown to be the total available area to one part in 10^{-6}.

As Birkhoff proved,[6] each elliptic periodic point generically has elliptic periodic orbits that encircle it, and elliptic periodic orbits that encircle them, *ad infinitum*. This complicates matters, but we avoid this complication by restricting explicit calculations in this paper to resonances of the primary "class",[7] for which the central periodic orbits are rotational rather than vibrational. Our formulae, however, are equally valid for any class of resonance. We believe that with an appropriate convention, the regions occupied by resonances of the same class never overlap.

RESONANCES AND CHIMNEYS

Let T be an area-preserving map. We denote phase-space points by capital letters and their components by small letters, for example, $X = (x, p)$. Orbits are denoted by subscripted variables, for example, $X_{t+1} = TX_t$. The phase-space is assumed to be a cylinder or annulus, with the configuration or angle variable, x, of period unity. The momentum coordinate is p. An example that we shall use for calculations is the "standard map":

$$p_{t+1} = p_t - k/2\pi \sin(2\pi x_t),$$

$$x_{t+1} = x_t + p_{t+1},$$

which has a parameter k. The pictures will all be in "symmetry" coordinates for this map, which are defined as $\{x, y = p - k/4\pi \sin(2\pi x)\}$.

Resonances

We introduce the general theory of resonances with a particular example from the standard map. When $k = 0$, the orbits of the standard map are extremely simple: the coordinate p is a constant of motion. Each orbit rotates with a frequency equal to p. For

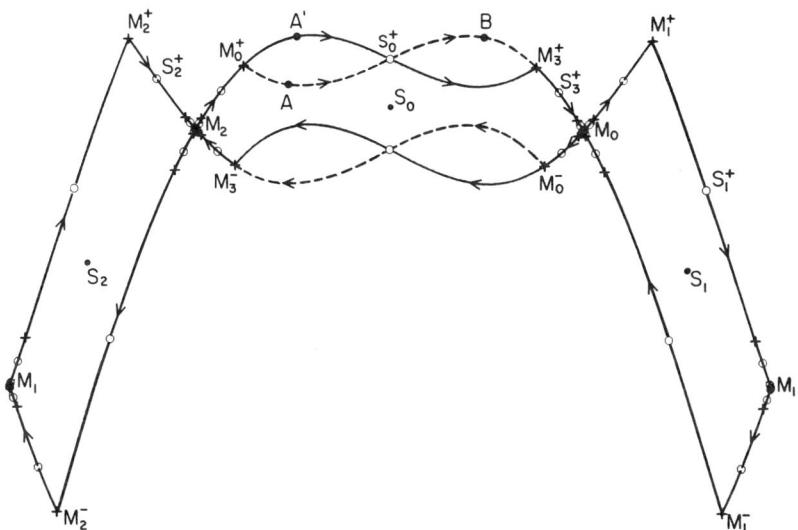

FIGURE 1. Resonance of frequency 1/3 for the standard map in symmetry coordinates. M_t are points on the period 3 hyperbolic orbit and S_t are points on the period 3 elliptic orbit. Orbits homoclinic to M_t are labeled by "+" on the upper separatrix and "−" on the lower separatrix. The resonance boundary is indicated by solid lines. Dashed lines represent the other boundaries forming the turnstiles in the partial separatrices.

nonzero k, the invariant circles of orbits with rational frequencies are destroyed. The Poincaré-Birkhoff theorem implies that there are at least two periodic points with each period. FIGURE 1 shows the 1/3 resonance, which is built on the framework of the two orbits of frequency 1/3.

In the figure, the points labeled M_t, $t = 0, 1, 2$, are the points on an orbit with negative "index". The index is defined as the number of rotations that the vector $(p', x') - (p, x)$ makes as the point (p, x) moves on a curve encircling the orbit. It is easy to see that a negative index orbit is unstable or hyperbolic. The points labeled S_t, $t = 0, 1, 2$, are a frequency 1/3 orbit with positive index; it is either a stable elliptic orbit or an unstable orbit that is hyperbolic with reflection; for present purposes, it does not matter which. (The symbols M and S actually stand for minimizing and saddle, thus representing properties of the action function.[2])

For an integrable system, there would be a separatrix joining the points M_t to form a chain of islands, with each island surrounding a point S_t. However, for nonintegrable systems, there is only a "partial separatrix", whose structure is a bit more complicated. There is an upper partial separatrix and a lower partial separatrix (labeled in FIGURE 1 by superscripts "+" and "−", respectively). The solid lines represent the boundary of the 1/3 resonance, and the dotted lines represent the remaining boundaries of the turnstiles through which the phase points move in and out of the resonance. The arrows represent asymptotic behavior.

The upper partial separatrix is constructed from the upper unstable and stable manifolds of the points of M_t. These manifolds always exist for hyperbolic periodic orbits. Poincaré showed that while the two manifolds do not necessarily coincide, area preservation implies that they at least must intersect. Choose any such point of intersection, say M_3^+ in FIGURE 1. The choice is arbitrary (see discussion below). The orbit of such an intersection point, shown as the +'s in the figure, is homoclinic to the orbit of M_0: it approaches this periodic orbit both in the past and in the future. To construct the upper resonance boundary, follow the right-going branch of the unstable manifold of M_2 until it reaches M_3^+. At M_3^+, switch to the stable manifold and follow it to M_0. This forms a segment of the upper partial separatrix; the remaining segments are formed from its preimages. The first preimage connects M_1 to M_2 and the second one connects M_0 to M_1. Thus, after two preimages, we obtain the upper solid curve shown in FIGURE 1, which is the upper resonance boundary. Of course, this boundary is not an invariant manifold: upon iteration, most of the boundary is unchanged; however, its preimage in the gap between M_2 and M_0 acquires a longer piece of stable manifold (the dotted segment in FIGURE 1), which is joined to a shorter piece of unstable manifold at M_0^+. Area preservation implies that there is another intersection point of the stable and unstable manifolds, S_0^+, on the dotted segment; the orbit of this point is also homoclinic to the period 3 orbit and is shown by the circles in the figure.

The region bounded by the figure eight formed by the manifolds connecting M_0^+ to M_3^+ is the upper turnstile of the resonance. It is easy to see from the construction that the left lobe of the turnstile is the set of points that, though currently below the partial separatrix, will be above it after one iteration. Similarly, the right lobe is the set of points that will cross from above to below on the next iteration. Thus, the turnstile is a doorway—and the only one—connecting the resonance to points above. An orbit below the partial separatrix must remain below until it lands in the left lobe of the turnstile.

The lower partial separatrix has a similar construction and properties, using lower stable and unstable manifolds. The resonance is the region bounded by the upper and lower separatrices. In FIGURE 1, it consists of the three curvilinear "rectangles" delineated by solid curves. They all have equal area. In order to leave the resonance, an orbit must land in the left lobe of the upper turnstile or the right lobe of the lower turnstile.

Chimneys

There is a certain amount of freedom in defining the resonance boundary that corresponds to the choice of homoclinic point where we switch from unstable to stable manifold. This does not matter too much when dealing with a single resonance. Indeed, the area of a resonance and the areas of the upper and lower turnstiles are independent of these choices. The only thing that changes is the time at which we decide that a trajectory leaves or enters the resonance. However, the choice is important when it comes to fitting resonances of different rotation number together. We would like to make a choice such that resonances never overlap because we aim to get a partition of phase-space.

The standard map is particularly simple as there is a natural choice: the map has a reflection symmetry about the line $x = 0$. One finds numerically that every periodic orbit $\{S_t\}$ has one point on this symmetry line. Thus, it is natural to choose turnstiles that have their center points on this line. Then, the union of these turnstiles forms a "chimney": all vertical transport takes place in this region (see, e.g., FIGURE 8).

This structure is a result of the symmetry of the standard map. We expect it to generalize to the class of reversible maps, that is, those possessing a "reflection" S (an orientation reversing map with S^2 = identity) that conjugates the map to its inverse ($STS = T^{-1}$). They always seem to have a similar, "dominant" symmetry line on which each positive index orbit has a point. Thus, we can choose this point to be the center of the corresponding turnstile. Then, the turnstiles all line up along this dominant line and the resulting "chimney" spans this line.

We have no theory of chimneys that applies generally, but we believe that the idea can be extended with care to more difficult cases in which the structure of the turnstiles may be more complicated or in which there may be more than one independent turnstile in a cantorus or partial separatrix.

ILLUSTRATIONS

Evolution of a Resonance

FIGURES 2–6 show the 2/5 resonance of the standard map at various values of the parameter k. To construct these pictures, we begin by finding numerically the two 2/5 orbits of the standard map. To do this, one can use the symmetry of the map to search for an orbit that starts on one symmetry line, or, as we have done, one can use the action principle that generates the map and search for extrema of the action.[2]

The orbits homoclinic to the hyperbolic $m/n = 2/5$ orbit are also obtained by action minimization, using periodic orbits as an approximation. The two homoclinic orbits on each separatrix are shown in the figures with the symbols "+" and "O". The periodic orbits used have frequencies slightly larger than or slightly smaller than the frequency of the resonance, 2/5. For example, in FIGURE 3, the upper homoclinic orbit is approximated by the orbit with frequency $m_+/n_+ = 43/107 = (1 + j \times 2)/(2 + j \times 5)$ and the lower one is approximated by the orbit with frequency $m_-/n_- = 43/108 = (1 + j \times 2)/(3 + j \times 5)$, for $j = 21$. As we increase the value of j from 1, the resulting periodic orbits converge to the homoclinic orbit extremely rapidly, and on the scale of the figures, there is certainly no difference between the actual homoclinic orbit and the periodic approximation. We choose j so that the larger eigenvalue of the approximating orbit is on the order of 10^{10} because this is an appropriate measure of the convergence.[2]

To obtain approximate upper stable and unstable manifolds, choose a point x_0

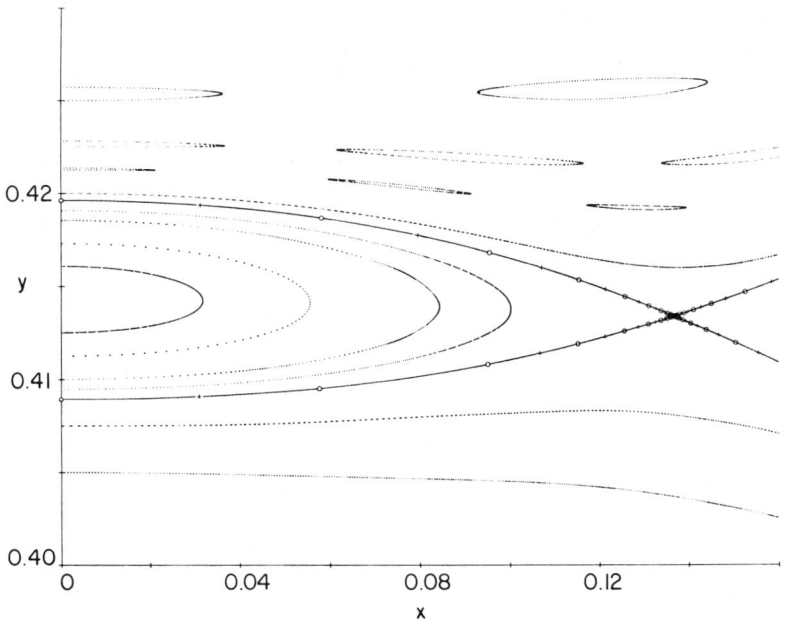

FIGURE 2. The 2/5 resonance of the standard map in symmetry coordinates at $k = 0.830$. The picture is reflection symmetric about $x = 0$. Homoclinic orbits are labeled with "+" and "O" on each separatrix. Most of the resonance is filled with invariant curves.

between the two points on the "+" m_+/n_+ homoclinic orbit closest to $x = 0$ (x_0 is in the "main" gap). Then, search for a value of y_0 that gives an orbit segment that returns exactly to x_0 after n_+ iterations and m_+ times around. If x_0 is not one of the points on the "O" or "+" homoclinic orbits, then y_n will not equal y_0: rather, the two points (x_0, y_0) and (x_0, y_n) are approximately on the stable and unstable manifolds in the main gap. The remaining points on the orbit of (x_0, y_0) fall in the other gaps of the m_+/n_+ orbit, the first $n_+/2$ points are approximately on the stable manifold, and the second $n_+/2$ points are approximately on the unstable manifold. As x_0 is varied over the main gap, the turnstile and separatrix are formed. Furthermore, as $m_+/n_+ \rightarrow m/n$ from above, these manifolds limit to the actual partial separatrix.

In FIGURE 2, $k = 0.830$ and the 2/5 resonance appears nearly integrable. The upper and lower separatrix turnstiles are so thin that they are not visible on the figure; in fact, the area of the lower turnstile is $\Delta W_- = 3.299 \times 10^{-8}$. Note, however, that the length of the turnstile is significant, approximately 0.03. This extreme aspect ratio of the turnstiles appears to be a typical phenomena and it also occurs in the case of turnstiles in cantori near the critical parameter value for breakup. The rate of convergence of the homoclinic orbit to the hyperbolic 2/5 orbit is governed by the unstable eigenvalue of that orbit, which is 1.92. Also shown in FIGURE 2 are four orbits trapped in the 2/5 resonance (which appear to lie on smooth invariant island curves), as well as three untrapped orbits (which appear to lie on invariant circles).

The parameter is increased to 1.025 in FIGURE 3 and the phase-space has become much more stochastic. The separatrix turnstiles are now visible, with $\Delta W_- = 1.35 \times 10^{-5}$, and the region bounded by the trapped invariant curves is significantly smaller than the resonance itself, which has an area of 1.62×10^{-2} (this is about 60% larger than at $k = 0.830$).

In FIGURE 4, the positive index 2/5 orbit has become unstable: it is now hyperbolic with reflection (its eigenvalues are negative reals). At the bifurcation that causes this instability, a stable period 10 orbit is born, which can be seen at the center of two island regions contained in each lobe of the 2/5 resonance. Of course, the bifurcation of the positive index orbit has not affected the stability or existence of the negative index orbit: it is still hyperbolic, and still has stable and unstable manifolds that comprise the partial separatrix. The area of the resonance continues to increase, now reaching 2.36×10^{-2}. Much of the region inside the resonance appears to be stochastic. Note also that the shape of the resonance is becoming increasingly different from the standard picture (e.g., the integrable pendulum); in fact, away from the symmetry line, the partial separatrices become nearly straight lines. This is due to the exponential contraction with the eigenvalue of the hyperbolic orbit. The sharp cusp in the resonance boundary is the point at which the change is made from unstable to stable manifold.

Just above $k = 1.375$, the period 10 orbit undergoes a period-doubling bifurcation itself. A slight further increase of k gives an infinity of period-doubling bifurcations that create period 5×2^n orbits. The bifurcations accumulate at $k \approx 1.42$. At the

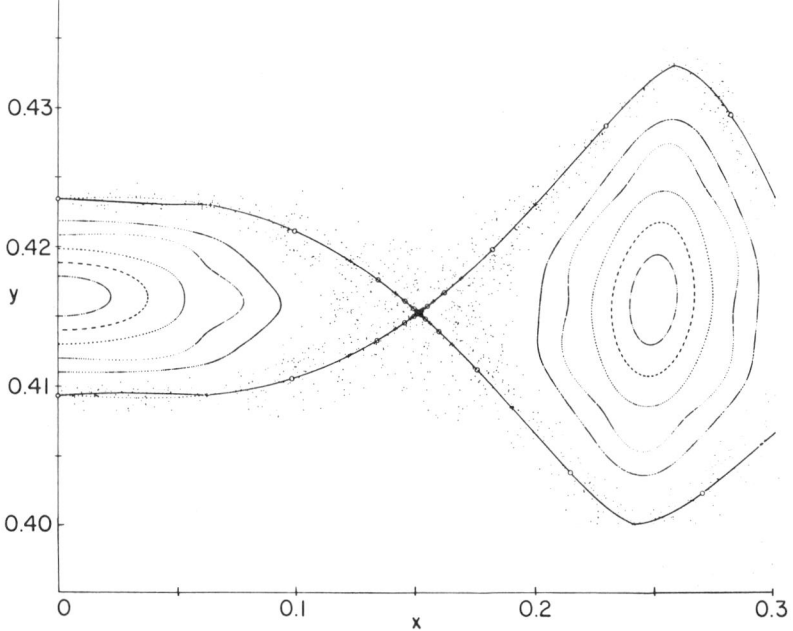

FIGURE 3. The 2/5 resonance at $k = 1.025$. Substantial stochastic regions are visible near the separatrix.

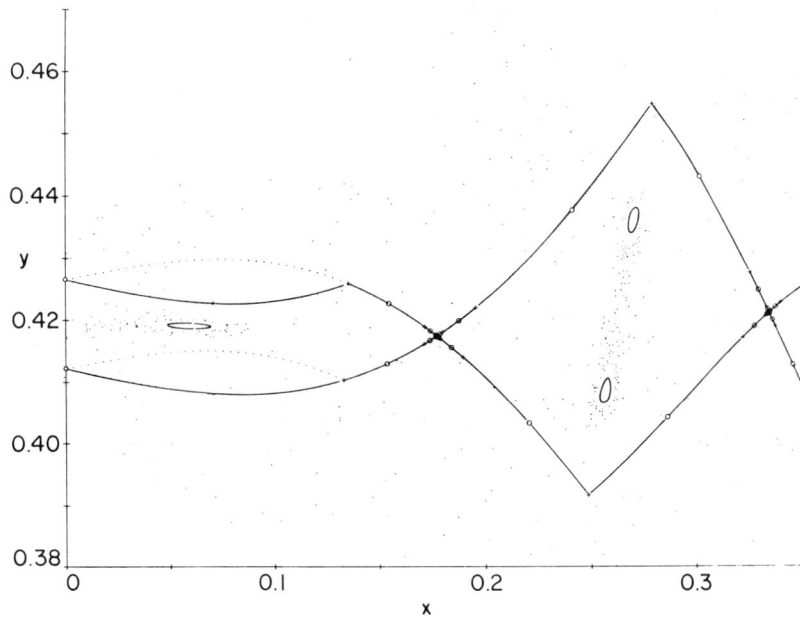

FIGURE 4. The 2/5 resonance at $k = 1.325$ after period doubling of the 2/5 $|_s$ orbit. Island regions around the period 10 orbit are shown.

accumulation point, all of the period 5×2^n orbits are unstable and, in fact, it is difficult to find any stable regions inside the resonance; however, it is possible that there are many extremely small elliptic regions.

By $k = 1.45$, in FIGURE 5, just after the accumulation of the period doublings, the resonance area reaches a maximum of 2.42×10^{-2}. The lower turnstile area has become $\Delta W_- = 1.30 \times 10^{-3}$, which is about 25% of that of one of the islands of the period 5 resonance chain. Roughly speaking, this implies that a point in the resonance has a probability of 0.25 of escaping from the resonance when it is in the island containing the turnstile. Of course, this occurs only once every five iterations because a point trapped in the resonance must rotate with the frequency of the resonance.

The subsequent evolution of the resonance is fairly mundane. The width of the island nearest the symmetry line increases monotonically. The remaining four islands cluster about the point $x = 0.5$, which is the minimum potential energy point. These four islands become increasingly rectangular in shape. The turnstile continues to grow monotonically with k. In fact, it is not difficult to show that for large k, every turnstile grows as $\Delta W \sim k/2\pi^2$. Because the resonance area decreases, this implies that the upper and lower turnstiles must at some point overlap (this first occurs at $k = 1.46$). Such behavior is shown in FIGURE 6, where $k = 1.75$. Here, the unstable manifold of the upper turnstile intersects the stable manifold of the lower turnstile. An orbit that falls in the overlap region will make a transition from above to below the resonance in one step. As k continues to increase, the overlap region becomes large; eventually, the turnstiles almost coincide and the probability of becoming trapped in the resonance is much smaller than that of skipping across it.

Area of a Resonance

The area of the 2/5 resonance as a function of the parameter is shown in FIGURE 7. For small k, area is increasing as one would expect from perturbation theory [a resonance with a frequency m/n appears at the n-th order in perturbation theory and area $\propto$ (resonant forcing)$^{1/2} \propto k^{n/2}$]. As we have seen, when k increases, the resonance shape alters dramatically and this estimate becomes invalid. When the eigenvalue of the hyperbolic orbit is approximately 6, its area reaches a maximum and begins to decrease as $k^{-\eta}$, where $\eta = 3.6$. In general, the value of η depends on the island frequency, but all resonances obey this law.

Tessellation of Phase-Space

When there are no rotational invariant circles (e.g., when k is larger than 0.972 for the standard map), the resonances nest together, and as we will see, they fill the entire phase-space. This is indicated in FIGURE 8, where both the 1/3 and the 2/5 resonances are shown. Note that between these resonances, there is a space that consists of exactly

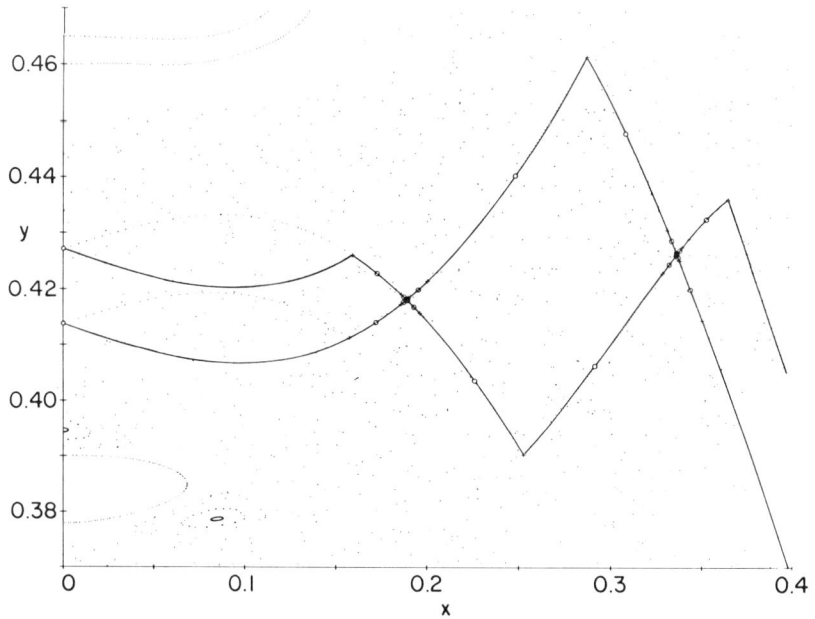

FIGURE 5. The 2/5 resonance at $k = 1.450$ when the resonance has reached its maximum area. The upper and lower turnstiles almost overlap.

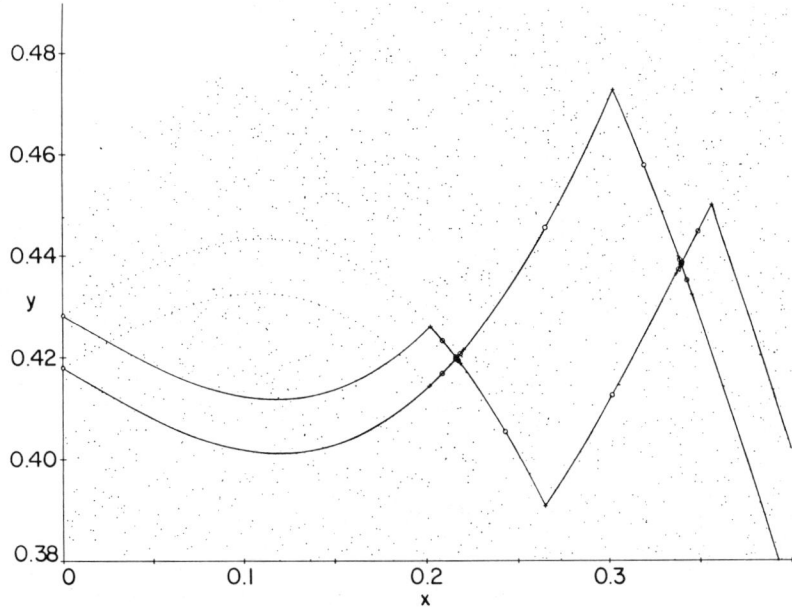

FIGURE 6. The 2/5 resonance at $k = 1.750$ where the upper and lower turnstiles have overlapped. The resonance interior in the gap around $x = 0$ is bounded by the solid curves. There is now substantial probability of crossing from above to below in one step.

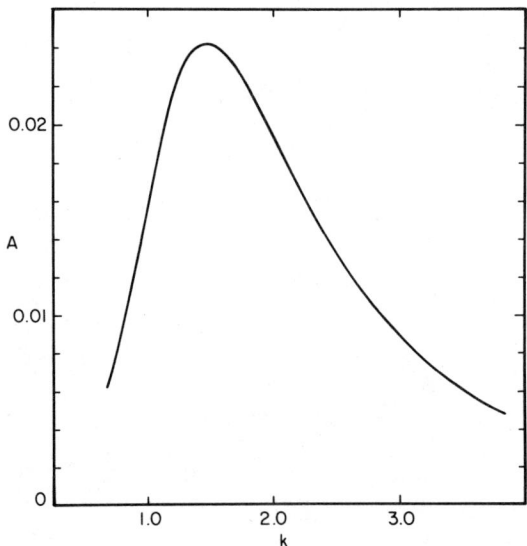

FIGURE 7. Area of the 2/5 resonance for the standard map as a function of the parameter k. The area reaches a maximum of 0.0242 at $k = 1.45$. At this point, the residue of the minimizing orbit is -2.08 and the positive index orbit is unstable. In fact, the period-doubling sequence of this orbit accumulates near $k = 1.42$.

eight rectangular regions. This number is no coincidence because the frequency 3/8 has the smallest denominator of any rational between 1/3 and 2/5. The 3/8 resonance is shown in FIGURE 9. It fits exactly between the previous two, thus leaving small rectangular gaps above and below that are to be filled by the next smallest denominator rationals, which are 4/11 and 5/13. This procedure continues and we obtain a tiling of phase-space with curvilinear rectangles.

To formalize this structure, one can use the Farey tree procedure to pick rationals in order of their denominator size. To obtain a Farey tree, define the zeroth generation by a pair of rationals, m_1/n_1 and m_2/n_2, satisfying $m_1 n_2 - m_2 n_1 = \pm 1$ (such rationals are called "neighboring"). A rational between two neighbors is obtained by

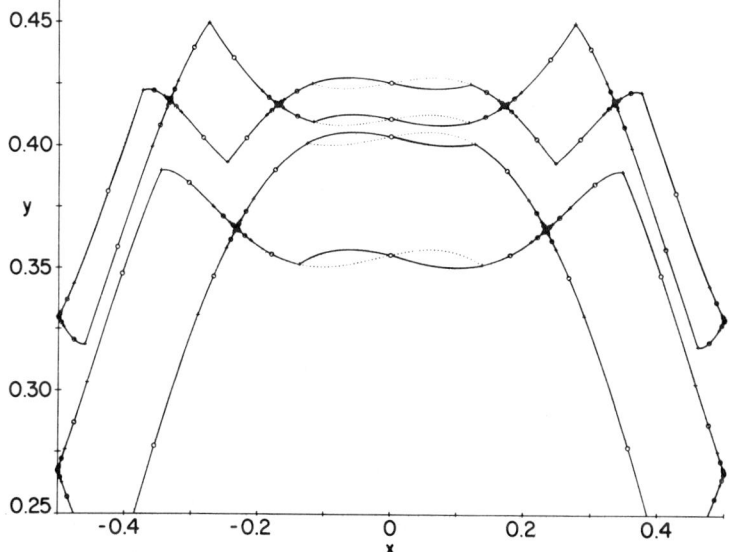

FIGURE 8. A neighboring pair of resonances (1/3 and 2/5) for the standard map in symmetry coordinates at $k = 1.25$

adding numerators and denominators: $m_3/n_3 = (m_1 + m_2)/(n_1 + n_2)$. This rational has the smallest denominator of any between the neighbors and, furthermore, it is a neighbor to each of its parents. Two new daughters can be constructed by again adding numerators and denominators of m_3/n_3 with each of its parents. By continuing this construction, one eventually obtains every rational in the interval $[m_1/n_1, m_2/n_2]$ exactly once.

The Farey tree of resonance areas is shown in FIGURE 10 (see reference 1 for details of the measurement of areas). Here, we include every resonance whose mother has an area larger than 10^{-6}. The area of all resonances between the 1/3 and 1/2 resonances is obtained by summing the area of each resonance on the Farey tree. Summing the

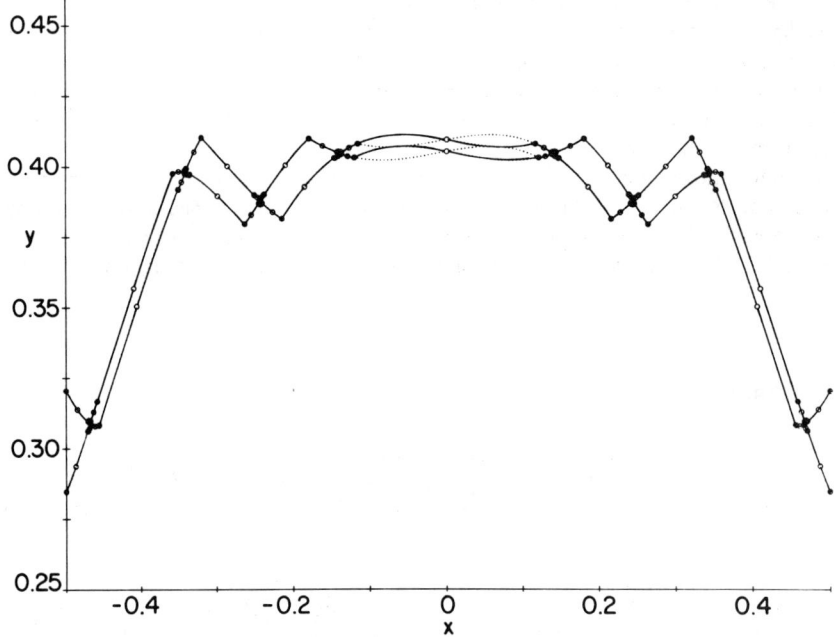

FIGURE 9. The 3/8 resonance at $k = 1.25$. Note that it fits neatly between the pair in FIGURE 8, leaving spaces for the next level on the Farey tree.

areas of all the resonances shown in FIGURE 10 gives the value

$$A_{\text{res}} = 6.63581 \times 10^{-2}.$$

The total area of the region of phase-space between the 1/3 and 1/2 resonances is given by the difference between the area below the lower partial separatrix of the 1/2 orbit and that below the upper partial separatrix of the 1/3 orbit. This gives

$$A_{\text{tot}} = 6.63582 \times 10^{-2},$$

which is only one part in 10^{-6} larger than A_{res}. We conclude that resonances occupy all of the phase-space area. This calculation has been done for other values of the parameter k. So long as $k > k_{\text{cr}}$, we find that resonances fill phase-space.

The resonance Farey tree is not geometrically self-similar. In fact, the decrease of log(*area*) along any path accelerates with increasing generation. This is due to the fact that for $k > k_{\text{cr}}$, no Farey paths approach frequencies that give invariant circles. Geometric scaling of areas should apply only for critical invariant circles. Indeed, we have previously found that the scaling near a cantorus proceeds as the exponential of an exponential,[1] and not geometrically, and thus we suspect that a similar result holds in this case.

CONCLUSIONS

Chaotic motion takes place within resonances, and a point remains within a given resonance until it reaches a turnstile, which is when it makes a transition to another resonance. Numerical results suggest that the resonances plus the rotational invariant circles form a partition of phase-space, up to a set of measure zero.

The resonance picture of chaotic motion is complementary to the picture based on partial barriers formed from cantori. The motion is seen to take place from resonance to resonance: resonances could be the states in a coarse-grained stochastic description. Resonances form a countable set, while cantori do not. This could lead to an orbit-coding scheme, where the code is a sequence of rationals designating which resonance an orbit is trapped in at which time. Our plan is to test numerically a Markov model based on resonances and to look for universal features of the resonance partition that enable one to extrapolate from calculations of a small number of resonances.

Below the value of k for the destruction of the last rotational invariant circle, one expects that the phase-space of the divided system is filled by the two components—invariant curves and resonances—with the area of invariant curves going smoothly to zero as k approaches k_{cr}. We have not verified this in detail because our technique for determining island areas by using periodic orbit approximations to the homoclinic orbits becomes more difficult as k decreases.

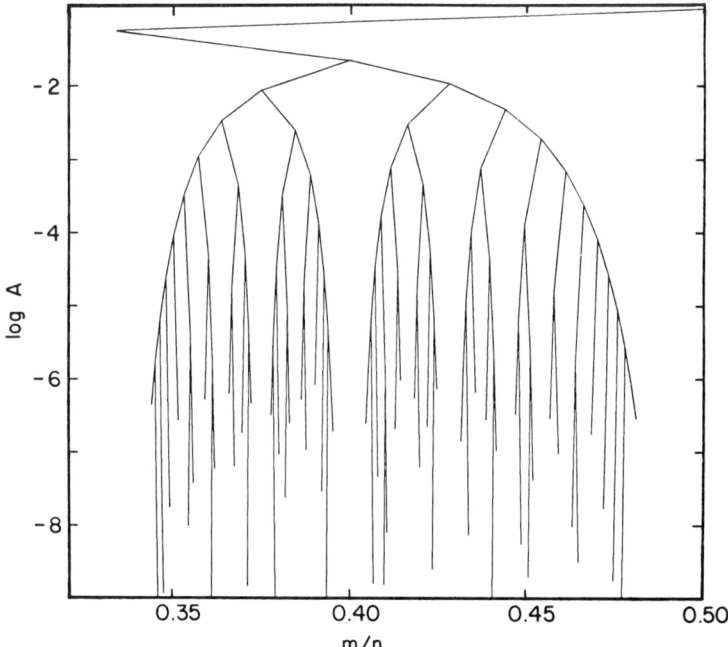

FIGURE 10. Resonance areas on the [1/3, 1/2] Farey tree for the standard map at $k = 1.283$.

The geometry of the resonances is not simple. There are resonances corresponding to every rational, with the area monotonically decreasing as one moves down the Farey tree. For reversible maps, where the largest gaps line up on a symmetry line, there is a chimney in which all transport occurs. The range of possible transitions and the size of the chimney depend on the details of the area-preserving map, but far from the chimney, a universal structure for the resonances is possible. The details remain to be worked out.

For larger numbers of degrees of freedom, there are in general no complete barriers to the chaotic motion. However, it may still be possible to define resonances, and transitions between resonances, so the present picture opens up the possibility of treating transport in systems of arbitrary number of degrees of freedom on a common basis.

ACKNOWLEDGMENTS

This discussion reports the results of collaborative research with R. S. MacKay and I. C. Percival.

REFERENCES

1. MACKAY, R. S., J. D. MEISS & I. C. PERCIVAL. 1984. Phys. Rev. Lett. **52:** 697; Physica **13D:** 55.
2. MACKAY, R. S., J. D. MEISS & I. C. PERCIVAL. 1987. Resonances in area preserving maps. Physica **25D**. In press.
3. CHIRIKOV, B. V. 1979. Phys. Rep. **52:** 265.
4. CHANNON, S. R. & J. L. LEBOWITZ. 1980. Ann. N.Y. Acad. Sci. **357:** 108.
5. BARTLETT, J. H. 1982. Celest. Mech. **28:** 295.
6. BIRKHOFF, G. D. 1935. Mem. Pont. Acad. Sci. Novi Lyncaei **1:** 85. Reprinted as: 1950. Collected Mathematical Papers. Vol II: 530. Amer. Math. Soc. New York.
7. MEISS, J. D. 1986. Phys. Rev. **A34:** 2375.

Hamiltonian and Dissipative Chaos[a]

GEORGE SCHMIDT

Department of Physics
Stevens Institute of Technology
Hoboken, New Jersey 07030

INTRODUCTION

Hamiltonian systems of low dimensionality are by now well understood. A two–degree of freedom system, with a Hamiltonian $H(q_1, q_2, p_1, p_2)$, is energy conserving, so the phase-space has three relevant dimensions. Similarly, a time dependent one-dimensional system, $H(q, p, t)$, has trajectories in the (q, p, t) three-space. Instead of looking at trajectories in phase-space, it is convenient to consider the successive intersections of a trajectory with a two-dimensional surface (e.g., a plane). These intersections, first introduced by Poincaré, represent a map of the surface, where each point is sequentially mapped into other points on the surface. This surface is also called the surface of section. It can be proven that this surface of section map is area preserving.[1,2]

Each physical system, of course, has its own map, but they all share a number of characteristics. One frequently used map is the standard map,[3-6]

$$x' = x + y + K/2\pi \sin 2\pi x = x + y',$$
$$y' = y + K/2\pi \sin 2\pi x, \qquad (1)$$

where x and y are position and momentum coordinates and K is a parameter. Note that equation 1 explicitly carries (x, y) into (x', y') without the need to integrate along the trajectory between intersections with the surface of sections. This map describes, for example, the motion of a particle in an infinity of periodic potentials, with each one moving with a different phase velocity.[5] Time dependence is characterized by a sequence of δ functions; hence, the elimination of trajectory integrals.

This map is easy to compute and a few trajectories are shown in FIGURE 1. One simply selects a few initial x, y coordinates and lets the computer print out successive iterations. Because the map is periodic in x and y, it suffices to show $0 \leq x \leq 1$, $0 \leq y \leq 1$. In FIGURE 1, $K = 0.95$ was chosen. It exhibits the basic features of Hamiltonian maps: stable fixed points and periodic orbits, chaotic regions generated in the vicinity of unstable periodic orbits, and continuous curves. These curves are the Kolmogorov-Arnold-Moser (KAM) trajectories. There are periodic orbits of all periods and the KAM curves may be viewed as stable periodic orbits of infinite period. The importance of KAM curves is due to the fact that no trajectory can cross them. Hence, chaotic regions are bounded by KAM curves. Stable periodic orbits of finite length are surrounded by islands that are separated from chaotic regions by KAM curves. Chaos is characterized by the fact that adjacent initial conditions lead to

---

[a]Partial support for this work was provided by DOE Contract No. DE-AC02-84 ER 13146.

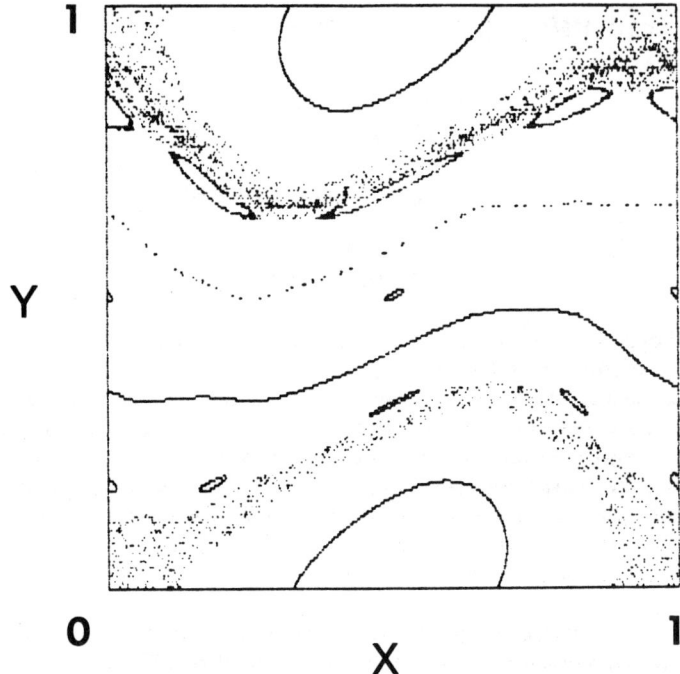

FIGURE 1. Some trajectories of the standard map of equation 1 with $K = 0.95$.

exponentially diverging trajectories. The exponent describing this divergence is the Lyapunov exponent.

As the strength parameter K is increased, the map changes. Stable periodic orbits destabilize and so do KAM curves, thus permitting chaos to spread and occupy larger areas. The stability properties of each periodic orbit can be calculated (hence that of KAM curves as well) as limits of periodic orbits of long period.[4-6]

The destabilization of periodic orbits typically takes place via period doubling. As K is increased, one finds that as a given orbit of period m destabilizes, a period $2m$ stable orbit is created at some K_1; this then becomes unstable at K_2, giving rise to a stable $2^2 m$ orbit, and so on. The values $K_1, K_2, \ldots, K_n, \ldots$ converge rapidly and in the limit

$$\lim_{n \to \infty} (K_n - K_{n-1})/(K_{n+1} - K_n) = \delta_H = 8.72 \ldots, \qquad (2)$$

which is a universal number for a large class of orbits.[7] At some finite $K = K_\infty$, there is no stable periodic orbit left in the vicinity of the formerly stable period m orbit.

THE DISSIPATIVE STANDARD MAP

Many systems of interest in physics and astrophysics have some dissipation. The flow in phase-space is no longer volume preserving, and neither is the map area preserving. In fact, the area shrinks upon successive mappings. The quantity characterizing dissipation is the Jacobian determinant of the map. For instance, one may extend the standard map by writing

$$x' = x + y',$$
$$y' = Jy + K/2\pi \sin 2\pi x, \qquad (3)$$

where J is the Jacobian.[8] $J = 1$ leads to the Hamiltonian map of equation 1, while for very large dissipation, one may set $J = 0$ and thus obtain the one-dimensional map

$$x' = x + K/2\pi \sin 2\pi x. \qquad (4)$$

If one runs the dissipative standard map (equation 3) on the computer with a small dissipation of $J = 0.99$, one obtains FIGURE 2. The two trajectories shown were started at $x = 0$, $y = 0.2$ and $y = 0.4$ (again for $K = 0.95$). After some chaotic looking transient, the first sequence converges to the fixed point $x = \frac{1}{2}$, $y = 0$, while the second converges to a period three stable orbit. Other initial conditions lead to similar results, that is, convergence to some stable periodic orbit. Careful calculations with other maps, for example, the Fermi map, lead to similar conclusions.[9] This result is similar to that found in integrable systems; a slightly damped pendulum finally approaches an equilibrium position corresponding to a stable attractor in the phase plane.

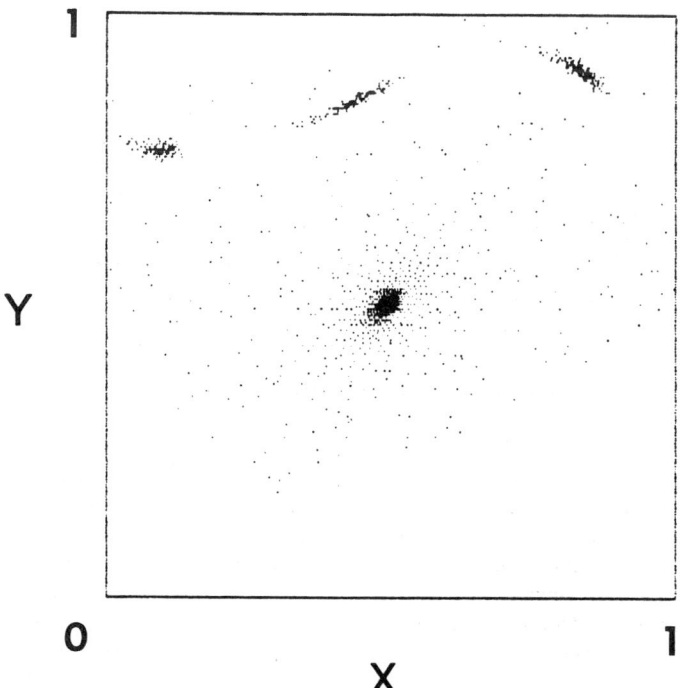

FIGURE 2. Two trajectories of the dissipative standard map of equation 3 with $K = 0.95$ and $J = 0.99$.

One might be tempted to conclude that dissipation kills chaos if one waits long enough. This is, however, contradicted by several observations. It is well known that a class of dissipative systems is chaotic due to a contraction of the surface of section plot to a strange attractor, such as the Hénon attractor.[10] Furthermore, in the $J = 0$ limit, the one-dimensional map of equation 4 exhibits a chaotic solution for a range of K values. FIGURE 3 shows the behavior of this map as a function of K. For K small, there is one stable attracting orbit, which goes through a period-doubling sequence. This is just like the one described in equation 2, except that now $\delta = \delta_F = 4.669\ldots$, which is the universal one-dimensional delta introduced by Feigenbaum.[11] At some K_∞, all

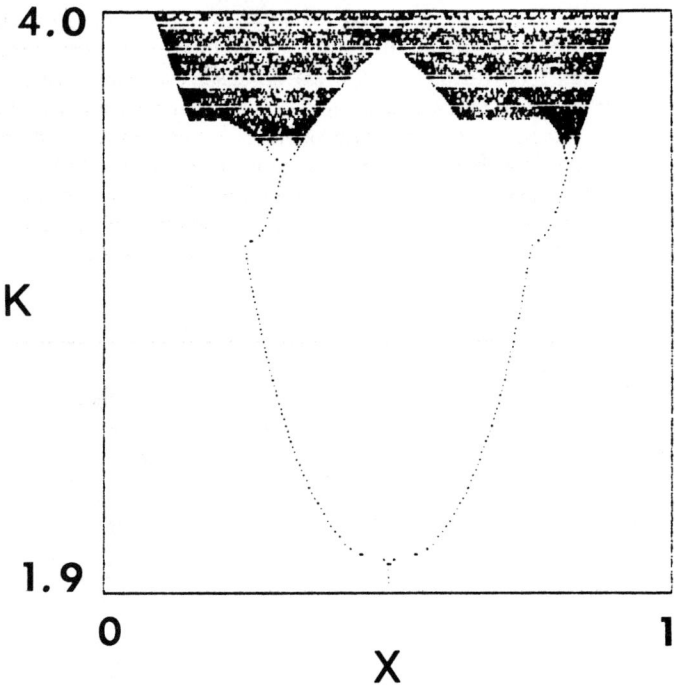

FIGURE 3. Bifurcation diagram of equation 4.

orbits are destabilized, and for $K > K_\infty$, chaotic bands with positive Lyapunov exponents emerge. These bands merge such that from 2^n bands, with increasing K, 2^{n-1} bands are produced.[12] This is the inverse bifurcation sequence, which is also characterized by δ_F. A complicating factor is the fact that the bands are interrupted by "windows" of stable nonchaotic orbits. For $J \neq 0$, one expects the chaotic bands to appear as strange attractors on the plane.

Because the Hamiltonian ($J = 1$) and the one-dimensional ($J = 0$) limits of equation 3 are well understood, it is natural to ask what happens to objects identified at the limits at intermediate values of J.

Consider, for instance, a stable periodic orbit on the Hamiltonian map. As K varies, one finds that it is created at some $K_{\min}(J = 1) = K_{\min}(1)$ by tangent bifurcation, and then it dies via period doubling at $K_{\max}(1)$. This periodic orbit therefore exists in the parameter range $K_{\min}(1) \leq K \leq K_{\max}(1)$. One may now construct analytically or computationally the $K_{\min}(J)$ and $K_{\max}(J)$ curves.[8] It turns out that these curves in parameter space extend all the way to $J = 0$. The range of existence, $\Delta K(J) = K_{\max}(J) - K_{\min}(J)$, becomes narrower with decreasing K and it ends up as one of the infinity of windows on the one-dimensional map. (The only exception is the period one fixed point, which ends up as the period one attractor.) Therefore, it appears that there is a one-to-one map in parameter space of the $\Delta K(1)$ ranges to the $\Delta K(0)$ ranges for all stable periodic orbits. This explains (at least qualitatively) why the ΔK ranges become narrower with smaller J. For $J = 1$ and a given K, a larger number of stable periodic orbits coexist in the x–y phase plane. At $J = 0$, there is only one stable orbit possible, so the windows in K must be narrow enough to accommodate the infinity of periodic orbits from the Hamiltonian map for any K.

Stable periodic orbits on the Hamiltonian map give elliptic orbits; that is, mapping an adjacent point results in sequences of points situated on ellipses around each image of the periodic orbit. With dissipation, these orbits become attractors; that is, adjacent points will asymptotically approach the orbit upon successive iterations. It is these attractors that swallow up the orbits in FIGURE 2.

An area A shrinks to $A' = JA$ after one mapping, to $A'' = J^2 A$ after two mappings, and to $A^{(m)} = J^m A$ after m mappings. Because a period m fixed point maps back to its starting position after m applications of the map, it is convenient to characterize it with the effective Jacobian $J_{\text{eff}} = J^m$. Even if the dissipation is small, $J = 1 - \epsilon$, for sufficiently long orbits (m large), the effective Jacobian, $J_{\text{eff}} = (1 - \epsilon)^m$, results in an arbitrarily large effective dissipation. In fact, one finds numerically that $\Delta K(J)$ shrinks very fast with increasing ϵ for longer orbits. We have seen that KAM trajectories are limits of periodic orbits as $m \to \infty$. Therefore, one expects $\Delta K(J)$ to shrink to zero for arbitrarily small dissipation for KAM trajectories.[8] This means that for a given K, one typically finds no remnant of the KAM trajectories in dissipative maps. This then enables the attractors to have finite basins and to attract faraway points (which, in fact, is what we have seen in FIGURE 2).

Next, we turn to the study of the chaotic bands of FIGURE 3 when the dissipation is reduced. Clearly, they can have no counterparts on the Hamiltonian map because 2^n piece chaos would imply that KAM lines bound each piece; however, there is no possibility to jump from one area to another across KAM boundaries. In Hamiltonian maps, only one piece chaos can exist.

One can follow numerically these orbits of 2^n piece chaos[8] starting at $J = 0$ for $0 \leq J < 1$. At $J = 0$, the two piece chaotic bands exist in the parameter range, $\Delta K_c^{(2)}(0)$. As J is increased, $\Delta K_c^{(2)}(J)$ decreases and disappears entirely at some $J = J_1$. The four piece band $\Delta K_c^4(J)$ behaves similarly and disappears at J_2, etc. One finds numerically that $J_n^2 = J_{n-1}$, so the extinction of 2^n piece strange attractors follows a well-defined law,[13] and when $J \to 1$, no 2^n piece chaos exists. Because the inverse bifurcation sequence producing the 2^n bands is universal (i.e., a large class of maps has this property), one is led to compute the ΔK_c ranges for other maps as well. FIGURE 4 shows the results, where the $K - K_\infty$ and J parameter plane is shown with the borders within which $2, 4, \ldots, 2^n$ piece chaos exists. These diagrams look similar for all maps studied[13] and they exhibit the $J_n^2 = J_{n-1}$ scaling. There seems to be a new law behind this, which will be explored using renormalization theory.

UNIVERSAL BEHAVIOR IN STRANGE ATTRACTORS

Just as the stable areas of phase-space can be understood by studying stable periodic orbits, chaos is associated with unstable ones. The 2^n piece strange attractor is produced by an unstable orbit of length 2^n, with an effective Jacobian $J_{\text{eff}} = J^{2^n}$. From $J_n^2 = J_{n-1}$, it follows that the effective Jacobians of all unstable periodic orbits are the same at the parameter values where the associated strange attractors disappear. We therefore expect some scaling law that depends on J_{eff}. In fact, it is known that for period-doubling bifurcations (relating to stable orbits), such scaling laws exist.[14,15]

The J_n values clearly accumulate at $J = 1$, and in this neighborhood, if $J_n = 1 - \epsilon_n$,

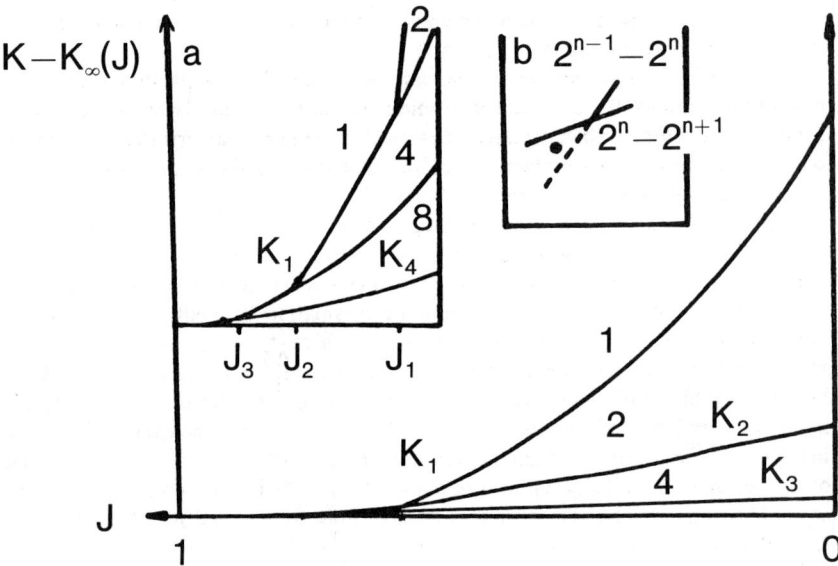

FIGURE 4. Location of inverse bifurcations in the $K - K_\infty$, J parameter space. The inset labeled "a" shows details of the region $0.6 < J < 1$.

the $J_n^2 = J_{n-1}$ relation gives

$$(1 - \epsilon_n)^2 = 1 - 2\epsilon_n = 1 - \epsilon_{n-1} \tag{5}$$

for small ϵ_n. The ϵ_n accumulate like a geometric series with the characteristic factor two, or

$$\lim_{n \to \infty} \frac{J_n - J_{n-1}}{J_{n+1} - J_n} = 2, \tag{6}$$

which is a law similar to that of the accumulation of the period-doubling bifurcation sequence given in equation 2.

The universal nature of period-doubling sequences, like the one given in equation 2, is the result of the following argument: There exists a universal function $g(x)$ that is unchanged on iteration when it is rescaled by an eigenvalue α:

$$g(x) = \alpha g[g(x/a)] \equiv R_F(g). \tag{7}$$

This expresses the self-similarity of period-doubling bifurcations upon iteration and rescaling.[11] The R_F is the renormalization operator associated with this process. In function space, the R_F is a mapping and the function $g(x)$ is a fixed point of this map. Stability can be ascertained by perturbing $g(x)$ and by looking for eigenvalues λ of the equation

$$R_F[g(x) + \epsilon f_i(x)] = g(x) + \lambda_i \epsilon_i f_i(x). \tag{8}$$

There are, of course, an infinity of eigenfunctions f_i and eigenvalues λ_i. With a few exceptions, it turns out that $|\lambda_i| \le 1$; that is, repeated iterations of the renormalization operation will attract these functions $g(x) + \epsilon f_i(x)$ to $g(x)$. A notable exception is the eigenvalue $\lambda_1 = \delta_F = 4.669\ldots$, which indicates that $g + \epsilon f_1$ will be repelled from g in function space with the repelling factor δ_F. Hence, the accumulation rate in K of period-doubling bifurcations appears here as an eigenvalue of the renormalization operator. In effect, setting $K = K_\infty$ in a map is equivalent to $f_1 = 0$, with the consequence that upon renormalization, $g(x)$ will be approached (aside of unimportant coordinate transformations corresponding to other $\lambda > 1$ values).

A similar scenario is known to exist for Hamiltonian maps. There exists a universal two-dimensional area-preserving map[7] T^* such that

$$R_H(T^*) = T^*, \tag{9}$$

where R_H is the Hamiltonian renormalization operator,

$$R_H(T) = B \circ T \circ T \circ B^{-1}. \tag{10}$$

Here,

$$B = \begin{pmatrix} \alpha & 0 \\ 0 & \beta \end{pmatrix} \tag{11}$$

is a two-by-two matrix, with α and β representing rescaling factors in x and y, where the x-axis is a symmetry line. Perturbation of T^* proceeds just as seen for $g(x)$. It leads again to an infinity of stable eigendirections, $|\lambda_i| \le 1$, and a few unstable ones containing the Hamiltonian bifurcation rate $\delta_H = 8.72\ldots$ (as well as those corresponding to unimportant coordinate transformations). The perturbed maps $T^* + \epsilon T_i$ are all area-preserving Hamiltonian maps.

Let us now enlarge function space to include dissipation as well.[13] One then finds that in addition to T_δ corresponding to a Hamiltonian map with eigenvalue $\lambda = \delta_H$, a new eigenvalue $\lambda = 2$ appears (as one anticipated from equation 6), which corresponds to a dissipative eigenperturbation T_J. Hence, dissipative maps in the vicinity of T^* are universal not only for period-doubling sequences, but for strange attractors as well. The sequential disappearance of the 2^n piece attractors is just one of many consequences of this result. In fact, considering the two-parameter function with dissipation $T^* + \epsilon_J T_J + \epsilon_\delta T_\delta$, the repeated application of the renormalization transformation will produce an infinite set of self-similar dissipative maps.[16] If ϵ_J, ϵ_δ were chosen such that the map would produce a 2^n piece strange attractor, then the renormalized map would give a similar 2^{n-1} piece attractor, and so on. All other aspects of the phase plane, like homoclinic tangencies, heteroclinic intersections, windows, etc., would all reproduce exactly.[16]

The renormalization calculation is nontrivial[13] and is only briefly outlined here. First, one generates T^*, which consists of two parts, $T_x^*(x, y)$ and $T_y^*(x, y)$, by choosing finite polynomials in x, y for both functions with undetermined coefficients. These polynomials are then inserted in equation 9 and the coefficients, as well as α and β, are determined by using a symbolic manipulation program on the computer. The length of the polynomials necessary to give a good approximation to T^* is determined by the convergence of the results as longer polynomials are chosen. Next, one adds polynomials to T^* as a perturbation, applies the renormalization operation, and looks for eigenvalues and the corresponding eigenfunctions, using again symbolic manipulation methods. The eigenvalue 2.000 . . . was determined to four significant figures.

STRUCTURE OF STRANGE ATTRACTORS

Chaos in dissipative systems is characterized by strange attractors. While chaotic regions in Hamiltonian surface of section maps occupy two-dimensional areas, the shrinking of areas in successive iterations in dissipative maps results asymptotically in chaotic objects of fractal dimensionality with dimensions less than two. These objects are the strange attractors. They are closely associated with the invariant manifolds of unstable periodic orbits.

FIGURE 5 shows one point of an unstable periodic orbit with part of its stable and unstable invariant manifolds. Successive iterations of a point on the stable manifold will map to other points on this manifold that approach the periodic orbit point asymptotically. Iterations of a point on an unstable manifold will produce points moving away from the orbit point in ever larger steps on this manifold. FIGURE 5 also illustrates the map of an area element in the vicinity of an unstable manifold. The length l grows into $l' = \lambda l$ (where $\lambda > 1$ is the unstable eigenvalue of the Jacobian determinant), while A shrinks into $A' = JA$. Evidently, all points in A will approach the unstable manifold itself upon a few iterations. Hence, it is important to follow the unstable manifold in its further path on the plane.

Beyond the end of the bifurcation sequence $(K > K_\infty)$, there are an infinity of periodic orbits, each with its own manifolds. This is illustrated schematically on FIGURE 6, where points of the $2^{n-1}, 2^n, 2^{n+1}$ orbits and their manifolds are shown. It is known[17] that the unstable manifold of any 2^n orbit always intersects the stable manifold of the 2^{n+1} orbit. Therefore, iterations of a point near the manifolds of the 2^n orbit may carry it to the stable manifold, and from there to the unstable manifold of the 2^{n+1} orbit. On FIGURE 6, the unstable manifold of the 2^n orbit is not connected with the stable manifold of the 2^{n-1} orbit. As K is increased, a heteroclinic crisis[18] takes place, as shown by the dashed line establishing this connection. There are now two options: one may pass from the manifold of the 2^n orbit to the manifolds of either the 2^{n+1} or 2^{n-1} orbit.

We illustrate this[13] with the "switch diagram" in FIGURE 7. The horizontal arrows indicate the path from the stable to the unstable manifolds of each orbit. These are always available, and so are the pathways from the unstable manifold of each orbit to the stable manifold of the "lower" one. In addition, for any $K > K_\infty$, an infinity of heteroclinic crises have taken place that connect lower unstable manifolds to higher stable ones.

It is useful to think of heteroclinic crises as "switches". As K is increased, higher switches are closed, thus making higher manifolds accessible from below. With the dashed switch open in FIGURE 7a, we have a 2^n piece strange attractor containing all lower ones as its parts. When the dashed switch is closed (at some higher K), a 2^{n-1} piece attractor results.

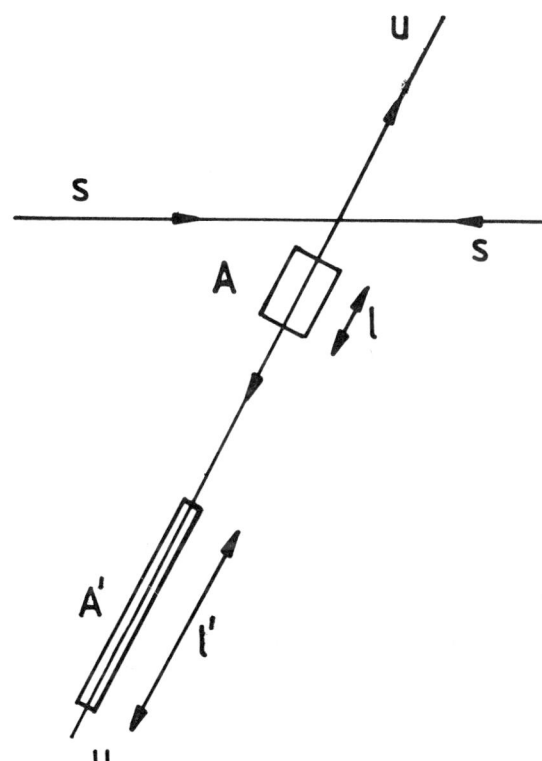

FIGURE 5. The vicinity of an unstable periodic point in a dissipative map with parts of the stable and unstable invariant manifolds. Area element A maps into $A' = JA$ and its length l maps into $l' = \lambda l$.

In parameter space, the lines shown on FIGURE 4 are the heteroclinic crisis lines where new switches are closed. The critical points J_n are those whose two switches close simultaneously. FIGURE 7a illustrates the situation for $J \ll 1$, where the sequence of crises (with increasing K) carries the system gradually to a one piece attractor. The situation for $J_n > J > J_{n+1}$ is shown in FIGURE 7b. In addition to the infinity of lower switches, a finite number of upper switches are also closed. The strange attractor corresponding to this state is a 2^{n+1} piece one. As the dashed switch is closed, a one piece attractor is suddenly obtained.

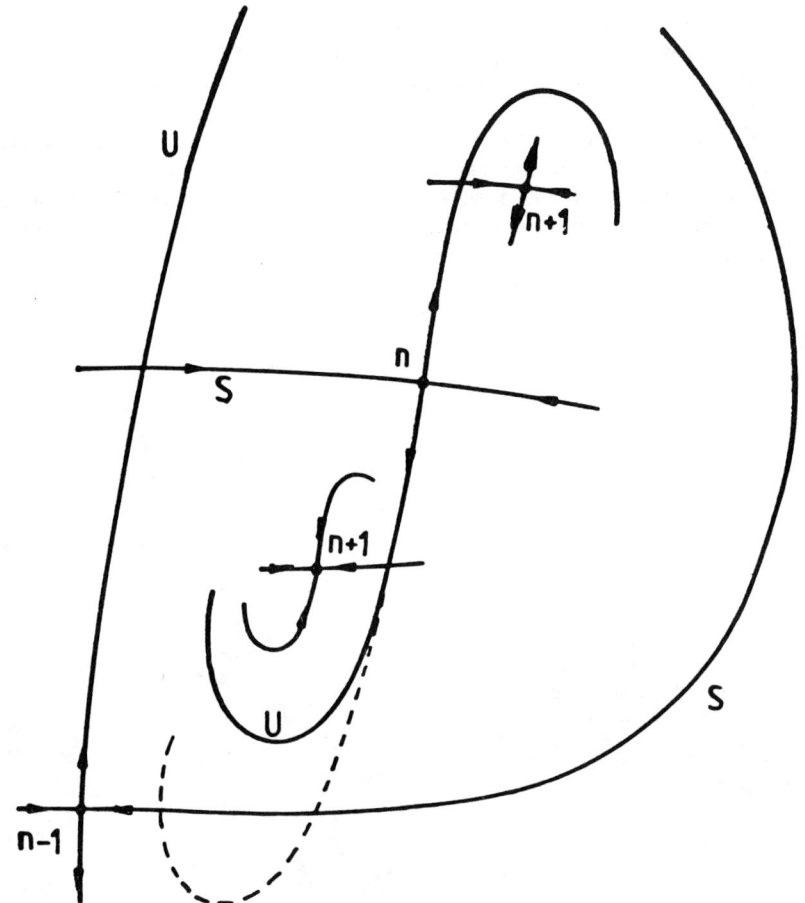

FIGURE 6. Schematic representation of stable (S) and unstable (U) manifolds of period 2^{n-1}, 2^n, and 2^{n+1} points. As K is increased, the intersection shown by the dashed line develops.

Two additional remarks are in order. First, as K is increased, there is a homoclinic crisis on the 2^n manifolds before the heteroclinic crisis connecting the 2^n and 2^{n+1} manifolds is reached. A homoclinic crisis produces intersections (an infinity of them) of the stable and unstable manifolds of the same orbit. This leads to a more complicated geometry of manifolds than the one shown in FIGURE 6, but it does not change the basic description of the phenomena.

The other remark is more subtle, and it relates to the last attractor seen as K is increased. In FIGURE 3, the last attractor shown is the one piece attractor. In the universal map, or the quadratic Hénon-type map,[7] which we write as

$$x' = -Jy + f(K, x),$$
$$y' = x - f(K, x'), \qquad (12)$$

where f is a quadratic map in x, the situation is different. For such maps, there is an attractor at infinity, and a further crisis is observed when the switch between this attractor and the rest is closed. The critical $J = J_0$ value corresponding to this transition fits into the universal scenario as $J_0 = J_1^2$. In these maps, in the Hamiltonian $(J = 1)$ limit, for $K = K_\infty + \epsilon$, points will wander off toward infinity following the stable

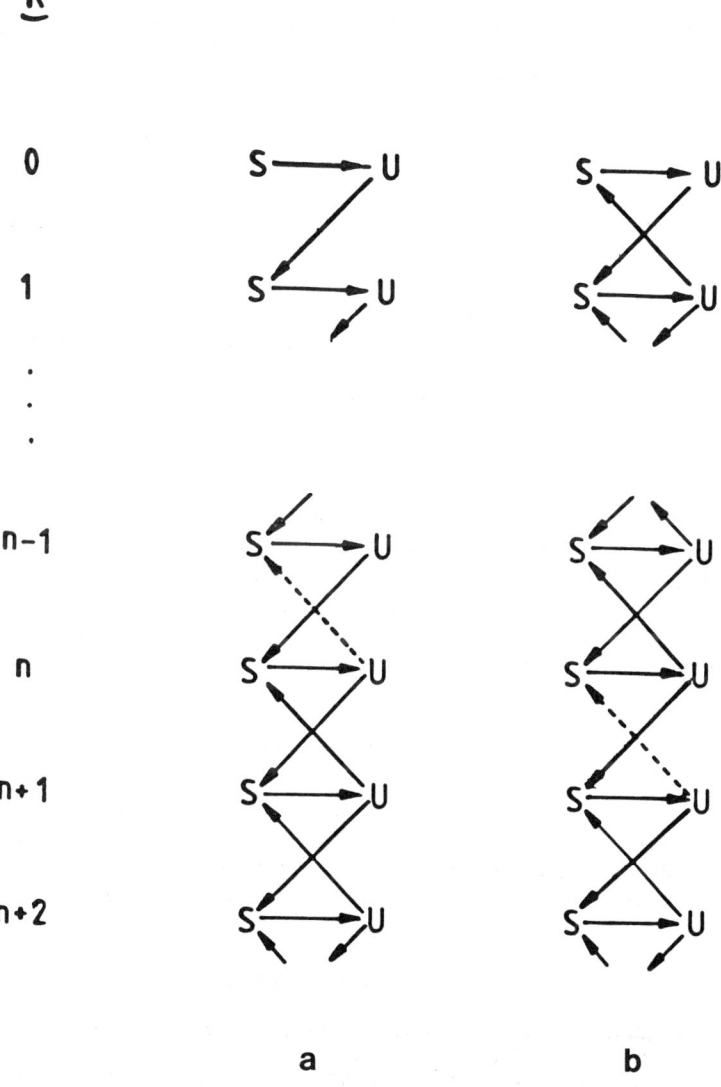

FIGURE 7. Switch diagram representing the flow between manifolds of 2^n, $K = 0, 1, 2, \ldots$, periodic orbits. Column "a" shows attractor dynamics for a small Jacobian, while column "b" is for a value slightly below some critical J_n.

manifold of the fixed point at infinity. There is, in effect, a crisis connecting the invariant manifold to that at infinity.

CHAOS IN DISSIPATIVE SYSTEMS

In a real physical system (e.g., the standard map), there are an infinity of periodic orbits in the Hamiltonian limit. As K is increased, one expects each one of these to undergo the universal bifurcation scenario. Of course, each one of these orbits has its own K_∞. When $K = K_\infty + \epsilon$ for any of these orbits, a crisis ensues that connects the invariant manifold not to infinity, but to the invariant manifold of some other orbit. The chaotic region seen in FIGURE 1 is the composite of an infinity of manifolds corresponding to the infinity of periodic orbits, each one above its own K_∞ located in the chaotic region. The stable orbits are located in islands and are shielded from each other, as well as from the chaotic trajectories, by KAM lines.

In dissipative systems, as we have seen, there are no KAM lines, so the shielding is not operative. As K is increased, a given periodic orbit bifurcates to its K_∞ and then goes through inverse bifurcations that terminate in some unstable orbit depending on J as in the universal map. The last crisis (the one that takes it to infinity in the universal map) will connect it to some other attractor. This attractor can be either a strange attractor, in which case chaos persists in this region, or a regular attractor, in which case local chaos is wiped out. In FIGURE 2, there are several powerful regular attractors in phase-space, with large basins of attraction. There still are many very-small chaotic regions with strange attractors, but most points will be attracted to regular attractors like the period one and period three attractors in FIGURE 2. For the parameters of this figure, we may say that chaos is practically nonexistent. However, as K is increased further so that the remaining stable orbits become unstable, the strange attractors of different period orbits link up into a large composite strange attractor and most of the phase-space becomes chaotic. It would be useful to have a quantitative measure of regular versus strange attractor basins as a function of K, but at the present time this problem has not yet been solved.

ACKNOWLEDGMENTS

The bulk of the work reported here has been carried out in collaboration with C. Chen, G. Gyorgyi, and B. Wang. The encouragement by O. Manley is also very much appreciated. Last, but not least, the author wishes to thank the organizers of the Florida workshop, in particular, G. Contopoulos, for their hospitality.

REFERENCES

1. BERRY, M. V. 1978. In Topics of Nonlinear Dynamics. Vol. 46. S. Jorna, Ed.: 16. Amer. Inst. Phys. Conf. Proc. AIP. New York.
2. LICHTENBERG, A. J. & M. A. LIEBERMAN. 1983. Regular and Stochastic Motion. Springer Pub. New York.
3. CHIRIKOV, B. V. 1979. Phys. Rep. **52:** 263.
4. GREENE, J. M. 1979. J. Math. Phys. **20:** 1183.

5. SCHMIDT, G. 1980. Phys. Rev. **A22:** 2849.
6. SCHMIDT, G. & J. BIALEK. 1982. Physica **5D:** 397.
7. GREENE, J. M., R. S. MACKAY, F. VIVALDI & M. J. FEIGENBAUM. 1981. Physica **3D:** 468.
8. SCHMIDT, G. & B. W. WANG. 1985. Phys. Rev. **A32:** 2994.
9. LIEBERMAN, M. A. & K. Y. TSANG. 1985. Phys. Rev. Lett. **55:** 908.
10. See, for example: SCHUSTER, H. G. 1984. Deterministic Chaos. Physica-Verlag. Würzburg, Federal Republic of Germany.
11. FEIGENBAUM, M. J. 1978. J. Stat. Phys. **19:** 25; 1979. J. Stat. Phys. **21:** 669; 1983. Physica **7D:** 16.
12. See for example: COLLET, P. & J. P. ECKMAN. 1980. Iterated Maps on the Interval as Dynamical Systems. Birkhäuser. Basel.
13. CHEN, C., G. GYORGYI & G. SCHMIDT. 1968. Phys. Rev. A: Rapid Commun. **A34:** 2568.
14. ZISOOK, A. B. 1981. Phys. Rev. **A24:** 1640.
15. QUISPEL, G. R. W. 1985. Phys. Rev. **A31:** 3924; Phys. Lett. **112A:** 353.
16. CHEN, C., G. GYORGYI & G. SCHMIDT. Phys. Rev. A. To appear.
17. HOLMES, P. & D. WHITLEY. 1984. Philos. Trans. R. Soc. London **A311:** 43.
18. GREBOGI, C., E. OTT & J. A. YORKE. 1983. Physica **7D:** 181.

The Breakdown of KAM Trajectories

D. BENSIMON[a] & L. P. KADANOFF[b]

[a] AT&T Bell Laboratories
Murray Hill, New Jersey 07974

[b] The James Franck Institute
Chicago, Illinois 60637

INTRODUCTION

Hamiltonian systems with two degrees of freedom, for example, two coupled oscillators, are the simplest class of conservative systems that may exhibit a nontrivial dynamics. In this introduction, we briefly review the phenomenology associated with that class of systems.

If the coupling between the two oscillators is linear, the system is integrable and the motion decouples into normal modes. These are best described in terms of action-angle variables:[1]

$$I_1(t) = I_1^0; \qquad I_2(t) = I_2^0;$$
$$\theta_1(t) = \omega_1(I_1^0)t + \theta_1^0; \qquad \theta_2(t) = \omega_2(I_2^0)t + \theta_2^0; \qquad (1)$$

because the energy is conserved, $E = E(I_1^0, I_2^0)$, the motion in phase-space is confined to a torus. If one considers a cross section of the torus (Poincaré section, see FIGURE 1), then the dynamics defines an area-preserving mapping of that section into itself:

$$\theta_{n+1} = \theta_n + \Omega(I_{n+1}) 2\pi,$$
$$I_{n+1} = I_n, \qquad (2)$$

where the winding number $\Omega(I_{n+1})$ is the ratio of the two normal modes frequencies, $\Omega = \omega_1/\omega_2$. Two types of motion are possible:

(1) Cyclic motion: If Ω is rational ($\Omega = p/q$), there exist an infinite number of different cyclic orbits. For any initial condition (θ_0, I_0), the sequence $\{\theta_n\}$ repeats itself with period q: $\theta_{n+q} = \theta_n$ [mod 2π].
(2) KAM curve: If Ω is irrational, the sequence $\{\theta_n\}$ never repeats, but the points θ_n fill out a curve Γ.

If the coupling between the two oscillators is nonlinear, the system is in general nonintegrable. According to the Poincaré-Birkhoff theorem,[2] only two cyclic orbits remain for every rational winding number ($\Omega = p/q$): an elliptic orbit that nearby points tend to cycle around and a hyperbolic one that nearby points are repelled from. Some KAM curves (irrational Ω) may still exist depending on the strength of the nonlinear coupling and their proximity to a rational orbit (a destabilizing factor). However, the most important qualitative and quantitative difference between the dynamics of integrable and nonintegrable systems is the existence in the latter of regions of stochasticity. These are regions, R^c, of finite area such that if x lies in R^c, then there is a subsequence x_{j_n} in R^c that approaches x. Because the dynamics is reversible, stochastic regions are bounded by KAM curves; see FIGURE 2.

However, as the strength of the nonlinear coupling is increased, more and more KAM curves disappear, that is, become discontinuous, cantor set–like. As they disappear, the barrier to the extension of chaos vanishes, becomes leaky, and the stochastic regions may thus grow in size.

The purpose of this discussion will be to describe how this may happen. Because the phenomenology described previously is very generic and quite independent of the form of the Hamiltonian and its nonlinearities, we will choose to study some particular area-preserving map, namely, the standard map:

$$T_k : (r, \theta) \to (r', \theta') = \left(r - \frac{k}{2\pi} \sin 2\pi\theta, r' + \theta \right). \qquad (3)$$

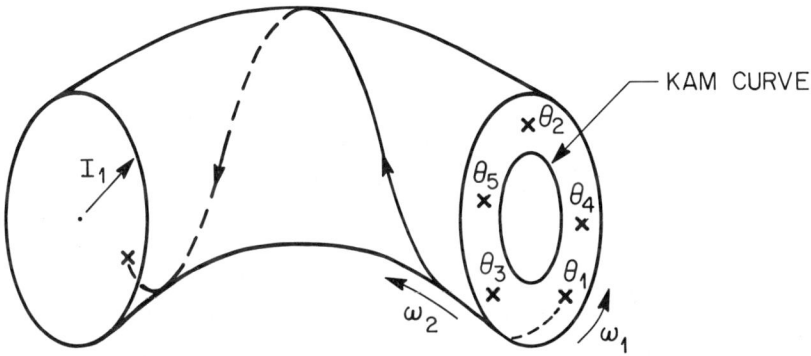

FIGURE 1. Phase-space dynamics for linearly coupled oscillators. The continuous curve is a KAM trajectory. The crosses are elements of an $\Omega = \tfrac{2}{5}$ cycle.

ESCAPE FROM A CHAOTIC REGION

As previously noted, stochastic regions are bounded by KAM curves. Thus, the action of the mapping T_k on a region R_Γ bounded by a KAM curve Γ is to map R_Γ onto itself:

$$T_k(R_\Gamma) = R_\Gamma. \qquad (4)$$

In order to understand the breakdown of Γ, one would like to construct a "good" estimate to it; this requires one to define what we mean by a "good" estimate. If we bound a region not by a KAM curve, but by an approximant of Γ, say γ, then

$$T_k(R_\gamma) \neq R_\gamma. \qquad (5)$$

The leakiness of γ can be measured by the area of $T_k(R_\gamma)$ that has escaped, that is, that overlaps with the complement of R_γ : $E - R_\gamma$. This leakiness is designated

$$L(T_k, R_\gamma) = S[T_k(R_\gamma) \cap E - R_\gamma]; \tag{6}$$

for a KAM curve, $L(T_k, R_\Gamma) = 0$. Thus, a "good" estimate γ to the KAM curve Γ is one that minimizes $L(T_k, R_\gamma) - (\delta L = 0 \text{ and } \delta^2 L < 0)$.

RELATING THE LEAKINESS TO THE ACTION

For specificity, we concern ourselves with the mapping (equation 3) defined on a cylinder, that is, with $\theta = \theta + 1$, and consider the KAM trajectory with winding number $\Omega = (\sqrt{5} - 1)/2$ (the Golden Mean), which is apparently the last one to disappear[1] at $k = k_c = 0.97163540...$.

Consider a rational approximant, $\Omega_n = p_n/q_n$, to Ω and construct a curve γ_0 passing through all the elements of the elliptic $\{(\theta_j^e, r_j^e)\}$ and hyperbolic $\{(\theta_j^h, r_j^h)\}$ cycles of

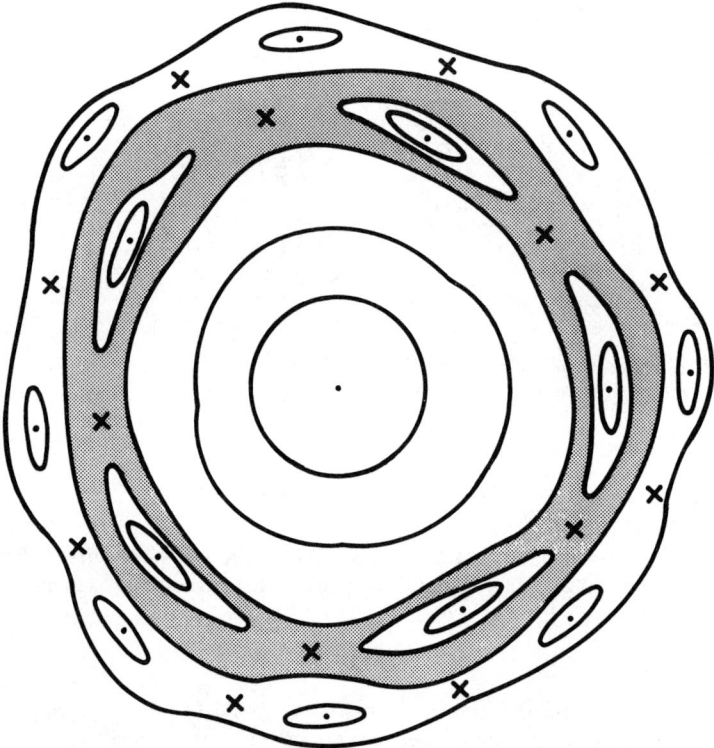

FIGURE 2. Phase-space dynamics for nonlinearly coupled oscillators. Dots: stable elliptic cycles. Crosses: unstable hyperbolic cycles. Continuous curves: KAM trajectories. Shaded area: bounded chaos. (Actually, the whole picture should be shaded because nonstochastic regions have measure zero.) This picture is repeated *ad infinitum* in the neighborhood of every elliptic cycle element.[1]

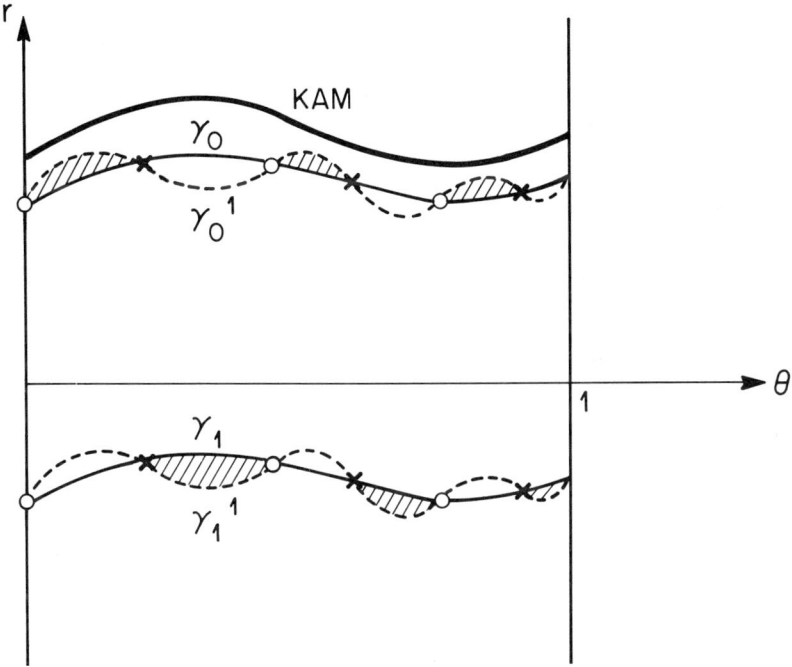

FIGURE 3. The leakiness of an approximant, γ, of the KAM curve, Γ, passing through all the elements of the elliptic and hyperbolic cycles of winding number $\Omega = \frac{2}{3}$. The escaped area is the dashed region.

winding number Ω_n. Let γ_1 be the same curve as γ_0, but displaced by one unit (see FIGURE 3). R_γ is the region bounded by γ_0 and γ_1. Notice that $T_k(\gamma)$ intersects γ at all of the elements of the cycles. Thus, the escaped area is

$$\frac{1}{2} L(T_k, R_\gamma) = \sum_j \int_{\theta_j^e}^{\theta_j^h} r \, d\theta - \int_{\theta_j^e}^{\theta_j^h} r' \, d\theta'. \tag{7}$$

Also, note that the motion is generated by an action principle in which

$$r' = -\frac{\partial}{\partial \theta'} A(\theta, \theta'), \qquad r = \frac{\partial}{\partial \theta} A(\theta, \theta'), \tag{8}$$

with

$$A(\theta, \theta') = -\frac{1}{2}(\theta - \theta')^2 - \frac{k}{(2\pi)^2} \cos 2\pi\theta.$$

Therefore,

$$\frac{1}{2} L(T_k, R_\gamma) = \sum_j \int_{\theta_j^e}^{\theta_j^h} \frac{\partial}{\partial \theta} A(\theta, \theta) \, d\theta + \int_{\theta_j^e}^{\theta_j^h} \frac{\partial}{\partial \theta} A(\theta, \theta') \, d\theta' = \sum_j A_h^q - A_e^q, \tag{9}$$

where A_h^q and A_e^q are the total actions for the hyperbolic and elliptic cycles, respectively, of length q, namely,

$$A_\alpha^q = \sum_{j=0}^{q} \left\{ -\frac{1}{2} (\theta_j^\alpha - \theta_{j+1}^\alpha)^2 - \frac{k}{(2\pi)^2} \cos 2\pi \theta_j^\alpha \right\}; \quad \alpha = e, h. \tag{10}$$

One can show[3] that this result is path independent as long as the original path considered, γ_0, passes through the cycle elements. Thus, γ_0 satisfies the first criterion for a good estimate of Γ: (L is stationary, $\delta L = 0$). However, it does not satisfy the second: γ_0 does not minimize L (actually $\delta^2 L = 0$). Nevertheless, it yields the correct scaling form for the leakiness, that is, the way L scales with q_n (as one gets a better and better estimate of Γ) and with $k - k_c$.

Thus, in order to study the disappearance of a KAM curve and its becoming leaky, one needs to study how the action behaves near $k = k_c$.[3,4] This can be done by a Renormalization Group (RG) analysis, whose principal ideas we discuss next.

RENORMALIZATION GROUP IDEAS FOR MAPPINGS

Presenting the scaling and RG analysis of the action for the Golden Mean KAM curve near its disappearance is beyond the scope of the present discussion and can be found in references 5–7. However, for the benefit of the neophyte, we will present and exemplify here the essential ideas behind the RG for mappings:

(1) Formulate an approximate problem and obtain a recursion relation between the structure of the problem, that is, some characteristic function at the level $n + 1$, to the structure at previous levels of the approximation.
(2) Formulate a scaling ansatz for the characteristic function.
(3) Let the approximation be exact, that is, let $n \to \infty$. Then solve the resulting fixed point equation and study its stability.

For example, consider the rational approximant to the Golden Mean KAM curve:

$$\frac{p_n}{q_n} = \frac{F_{n-1}}{F_n} \longrightarrow \frac{\sqrt{5}-1}{2} \equiv \Omega,$$

where the F_n's are Fibonacci numbers: $F_{n+1} = F_n = F_{n-1}, F_0 = F_1 = 1$. Let[6]

$$T_k: r' = r - \frac{k}{2\pi} \sin 2\pi\theta,$$

$$\theta' = r' + \theta,$$

$$R: r' = r,$$

$$\theta = \theta - 1. \tag{11}$$

Notice that $T_k \circ R = R \circ T_k$ [where "o" stands for the composition $f \circ g = f(g(x))$]. Let $z_n^0 = (r_n^0, 0)$ be a point belonging to a p_n/q_n cycle,

$$z_n^0 = G_n(z_n^0) \equiv T_k^{q_n} \circ R^{p_n}(z_n^0), \tag{12}$$

because

$$p_{n+1} = p_n + p_{n-1} \quad \text{and} \quad q_{n+1} = q_n + q_{n-1};$$

thus, by using the commutativity of R and T_k, one obtains the desired relation between the structure of the problem at level $n + 1$, that is, G_{n+1}, and its structure at levels n and $n - 1$:

$$G_{n+1}(z_0^{n+1}) = G_n \circ G_{n-1}(z_0^{n+1}) = G_{n-1} \circ G_n(z_0^{n+1}). \tag{13}$$

Next, make the scaling ansatz,

$$G_n = \tilde{\alpha}^{-n} g_n[\tilde{\alpha}^{-n}(s, \theta)], \tag{14}$$

with

$$\tilde{\alpha} = \begin{pmatrix} \beta & 0 \\ 0 & \alpha \end{pmatrix}$$

and $s = r - r^*$, where r^* is the value of the KAM curve at $\theta = 0$. Finally, let $n \to \infty$ ($g_n \to g^*$) and write the fixed point equation:

$$g^* = \tilde{\alpha} g^* \circ \tilde{\alpha} \circ g^* \circ \tilde{\alpha}^{-2} = \tilde{\alpha}^{-2} g^* \circ \tilde{\alpha}^{-1} \circ g^* \circ \tilde{\alpha}^{-1}. \tag{15}$$

Notice that g^* actually represents two functions. This then makes equation 15 harder to solve than a similar fixed point equation obtained for a scalar function such as the action.[5,7] Indeed, to our knowledge, equation 15 has not yet been solved. However, similar RG ideas can be applied to the action for the Golden Mean KAM curve near its disappearance.[5-7] They result in the following prediction for the scaling of the leakiness L (equation 9):

$$L(T_k, R_{q_n}) = q_n^{-do} L_{\gtrless}^*(q_n |k - k_c|^\nu), \tag{16}$$

with $do = 3.049960\ldots$, $\nu = 0.987463\ldots$, and where $L_{\gtrless}^*$ is a scaling function that applies respectively when $k > k_c$ and $k < k_c$. Its asymptotic behavior ($\xi \to \infty$) is

$$\begin{aligned} k < k_c : L_<^*(\xi) &\sim e^{-\gamma \xi} \quad (\gamma \text{ is a constant}), \\ k > k_c : L_>^* &\sim \xi^{do}. \end{aligned} \tag{17}$$

Namely, in the supercritical region, we expect

$$\lim_{q_n \to \infty} L(T_k, R_{q_n}) = |k - k_c|^{\nu do}. \tag{18}$$

This RG analysis using nonoptimal estimates of Γ was checked against numerical results with optimal estimates of Γ, that is, curves γ_n passing through all the elements of the hyperbolic cycle of length q_n, which can be shown to minimize the leakiness. The predictions of the RG calculation were thus confirmed.[3]

REFERENCES

1. LICHTENBERG, A. J. & M. A. LIEBERMAN. 1983. Regular and Stochastic Motion. Springer-Verlag. New York/Berlin.

2. BIRKHOFF, G. D. 1927. Dynamical Systems. Amer. Math. Soc. New York; see also reference 1.
3. BENSIMON, D. & L. P. KADANOFF. 1984. Physica **13D:** 82.
4. MACKAY, R. S., J. D. MEISS & I. C. PERCIVAL. 1984. Physica **13D:** 55.
5. KADANOFF, L. P. 1981. Phys. Rev. Lett. **47:** 1641.
6. SHENKER, S. J. & L. P. KADANOFF. 1982. J. Stat. Phys. **27:** 631.
7. MACKAY, R. S. 1983. Physica **7D:** 283; thesis (Princeton).

Fractal Basin Boundaries with Unique Dimension[a]

CELSO GREBOGI,[b] EDWARD OTT,[b,c] JAMES A. YORKE,[d,e]
AND HELENA E. NUSSE[f]

[b]*Laboratory for Plasma and Fusion Energy Studies*
[c]*Departments of Physics and Electrical Engineering*
[d]*Institute for Physical Science and Technology*
[e]*Department of Mathematics*
University of Maryland
College Park, Maryland 20742

[f]*State University of Groningen*
Groningen, the Netherlands

INTRODUCTION

It is common for nonlinear dynamical systems to have more than one stable time-asymptotic final state (i.e., more than one attractor). In these cases, the attractor to which an orbit tends will depend on its initial condition. The set of initial conditions that approach a given attractor is the basin of attraction for that attractor. A point is in the basin boundary if every ϵ-neighborhood of it contains points in at least two basins. Recently, there have been a number of papers investigating the properties of basin boundaries. In particular, basin boundaries have been shown to sometimes possess fractal structure;[1-4] thus, such basin boundaries are called fractal basin boundaries. In such cases, a convenient characterization of the basin boundary is in terms of the basin boundary dimension dim $(S \cap B)$, where S denotes some open set in phase-space that contains part of the basin boundary B (e.g., S might be a volume bounded by a smooth closed surface), and dim $(S \cap B)$ is the dimension of the part of the basin boundary that lies in S. The dimension of the basin boundary dim $(S \cap B)$ can be defined in different ways, and an appropriate choice of definition can often be made to reflect issues of practical interest in a given situation. This is discussed further in the next section.

Fractal basin boundaries can have important practical consequences. In particular, for the purposes of determining which attractor eventually captures a given orbit, the arbitrarily fine-scaled structure of fractal basin boundaries implies that sensitivity to small errors in initial conditions can be vastly increased as compared to the situation where the boundary is not fractal. The sense in which this is true is discussed in references 1 and 2, where it is also shown that the boundary dimension dim $(S \cap B)$ in S provides a quantitative measure of this sensitivity.

---

[a]This work was supported by DARPA under ACMP, the U.S. Department of Energy, Office of Basic Energy Sciences (Applied Mathematics Program), and by the Office of Naval Research.

The purpose of this paper is to show that in important special cases[g] (see examples of 1-D and 2-D maps in later sections of this paper), the dimension of basin boundaries can be unique. By unique, we mean that dim $(S \cap B)$ takes on the same value for all open sets S that contain part of the basin boundary. We ask, then, what property of the basin boundary insures uniqueness of its dimension (see the section entitled ASYMPTOTIC TRANSITIVITY ON THE BOUNDARY). To put this question in sharper focus, suppose P and Q are two points on the basin boundary. Trajectories starting on the boundary must stay on the boundary, but trajectories on the basin boundary will tend asymptotically to some subset of the basin boundary, usually not the whole basin boundary. Thus, none of the trajectories starting near P and Q may ever again come close to either P or Q.[h] Therefore, why should the dimension of the basin boundary near P be like that near Q? Why, as we deform and move S, do we not see the dimension dim $(S \cap B)$ varying through a range of values? The answer lies in understanding the long-term behavior of these trajectories. If there is a set Λ in the basin boundary such that the trajectories through P and Q come arbitrarily close to all points of Λ and these trajectories stay close to Λ, then the trajectories have similar long-term behavior. It is this property that makes the dimensions near P and Q equal. We formalize this idea in a later section with the discussion of "asymptotic transitivity on the boundary".

DIMENSION DEFINITIONS FOR BASIN BOUNDARIES

Box-Counting Dimension

The box-counting (or capacity) dimension is[1]

$$d_b(S \cap B) \equiv \lim_{\epsilon \to 0} \ln N(\epsilon, S)/\ln (1/\epsilon), \qquad (1a)$$

where $N(\epsilon, S)$ denotes the minimum number of D-dimensional cubes in a grid of edge-length ϵ needed to cover that part of the boundary that lies in S, and where D is the dimension of the phase-space. (Throughout this paper, we shall assume S to be an open set lying in a compact set on which the dynamics is defined.) It can also be shown[2] that $d_b(S \cap B)$ is also given by another equivalent definition,

$$d_b(S \cap B) = D - \lim_{\epsilon \to 0} \ln f(\epsilon, S)/\ln \epsilon, \qquad (1b)$$

where $f(\epsilon, S)$ is the fraction of the phase-space volume lying in S that is within ϵ of $B \cap S$.

Uncertainty Dimension

We now consider another definition of a basin boundary dimension. For example, say that we randomly pick an initial condition x with uniform probability in the volume

---

[g]These examples are nontrivial in that self-similarity is not built into the boundary structure. Self-similar examples obviously should have a unique dimension, but basin boundaries for typical systems are almost never self-similar.

[h]For example, this occurs for typical boundary points in the two-dimensional map example found later in this paper.

of an open subset of the phase-space S, where $S \cap B$ is not empty. Then, we pick another initial condition $y \in S$ randomly in the ball $|x - y| \leq \epsilon$. Let $p(\epsilon, S)$ be the probability that x and y are in different basins. We define the uncertainty dimension of B in S as

$$d_u(S \cap B) \equiv D - \lim_{\epsilon \to 0} \ln p(\epsilon, S)/\ln \epsilon. \qquad (2)$$

Thus, for fractal basin boundaries, we can have

$$p(\epsilon, S) \sim \epsilon^{\alpha_u(S)},$$

where $\alpha_u = D - d_u$, indicating sensitive dependence on an initial condition uncertainty[1,2] for $\alpha_u < 1$.

Comparison of d_b and d_u

The quantity $p(\epsilon, S)$ can be thought of as the probability of making an error in predicting which attractor a typical initial condition goes to. On the other hand, $f(\epsilon, S)$ can be thought of as the probability of choosing an initial condition such that it is *possible* to make an error in predicting to which attractor a typical initial condition goes. Thus, the latter probability is larger than the former,

$$f(\epsilon, S) > p(\epsilon, S),$$

and from equations 1b and 2, we have

$$d_b(S \cap B) \geq d_u(S \cap B). \qquad (3)$$

While there are many examples of fractal basin boundary sets for which $d_b = d_u$, it is also possible to construct examples for which $d_b \neq d_u$. Basin boundaries arising in typical dynamical systems, however, are a special class of boundary set and might therefore possess special properties. Indeed, based on examples, it has been conjectured that $d_b = d_u$ for basin boundaries in typical[i] dynamical systems.[2,5] No proof or counterexample to this conjecture currently exists. Our work in the following sections will be independent of whether or not this conjecture holds.

Properties of d_u and d_b

We wish to present here the following two properties of d_b and d_u:

(1) If $S_1 \supset S_2$ are two open sets containing part of the basin boundary, then

$$\dim(S_1 \cap B) \geq \dim(S_2 \cap B). \qquad (4)$$

(2) The dimensions $d_u(S \cap B)$ and $d_b(S \cap B)$ are invariant under smooth coordinate changes, $x' = g(x)$, where g is differentiable and invertible, and g^{-1} is differentiable on the closure of S.

---

[i]There is no appropriate way to rigorously express the idea of "almost all" dynamical systems, but by "typical", we mean that systems encountered in practical situations will be of this type.

(In subsequent sections, we shall write dim instead of d_b, d_u, and we shall often refer to "dimension" rather than to "box-counting dimension" or "uncertainty dimension". In such cases, the corresponding statement is meant to apply for both d_u and d_b. Thus, for example, equation 4 is meant to hold for both d_u and d_b.)

1-D MAPS WITH UNIQUE BOUNDARY DIMENSION

Fractal sets with self-similarity (e.g., the middle-third Cantor set) are well known and have a unique dimension precisely because of their self-similarity. However, self-similarity is rare in dynamics and, in general, does not occur for fractal basin boundaries. Here, we present the simplest kind of one-dimensional map for which there is a fractal basin boundary and we show that its dimension is unique, although it is not, in general, self-similar. The map, $x_{n+1} = f(x_n)$, is illustrated in FIGURE 1. [Self-similarity occurs for this map if $x_b - x_a = x_d - x_c$ and if $|f(x)|$ is the same constant in the intervals $(0, x_a)$, (x_b, x_c), $(x_d, 1)$; this, however, is not assumed and, typically, there is no self-similarity.] For this example, we assume that f' is nonzero in $[0, x_a] \cup [x_b, x_c] \cup [x_d, 1]$, and for almost all initial conditions, x_n tends to either $+\infty$ or $-\infty$ (we regard these as the two attractors). From the figure, $(-\infty, 0)$ is part of the $-\infty$ basin and $(1, +\infty)$ is part of the $+\infty$ basin. It remains to give the basin structure in $[0, 1]$. Let

$$\Sigma_k = \{x \in [0, 1] | f^i(x) \in [0, 1], \quad i = 1, 2, \ldots, k\}.$$

Notice that Σ_k is the union of 3^k intervals. The basin boundary is thus the Cantor set,

$$B = \bigcap_{k=1}^{\infty} \Sigma_k.$$

Proposition: The basin boundary C has a unique dimension.

Proof: For all intervals $S \supset [0, 1]$, the dimension $d(S)$ is the same, and we denote this dimension value as $\hat{d}$. Now, let S be any open interval containing part of C. Then, because we can choose another larger set S_+ that contains both C and S, equation 4 implies

$$d(S_+) = \hat{d} \geq d(S).$$

Also, because the interior of S contains part of C, then as we look at Σ_k for larger and larger k, eventually we must find that one of the 3^k components of Σ_k falls entirely within the interior of S. We denote such a component as σ_k. By construction, the k-th iteration of σ_k is the interval $[0, 1]$, $f^k(\sigma_k) = [0, 1]$. Thus, there is a smooth change of variables relating σ_k to $[0, 1]$, namely, f^k. Because the dimension is preserved under smooth variable changes with nonzero derivative, we have that

$$d(\sigma_k) = d([0, 1]),$$

which in turn is just $\hat{d}$. Because $S \supset \sigma_k$, equation 4 yields

$$d(S) \geq d(\sigma_k) = \hat{d}.$$

Combining the inequalities on $d(S)$, we have $d(S) = \hat{d}$ for all S containing part of C in its interior; that is, the dimension is unique.

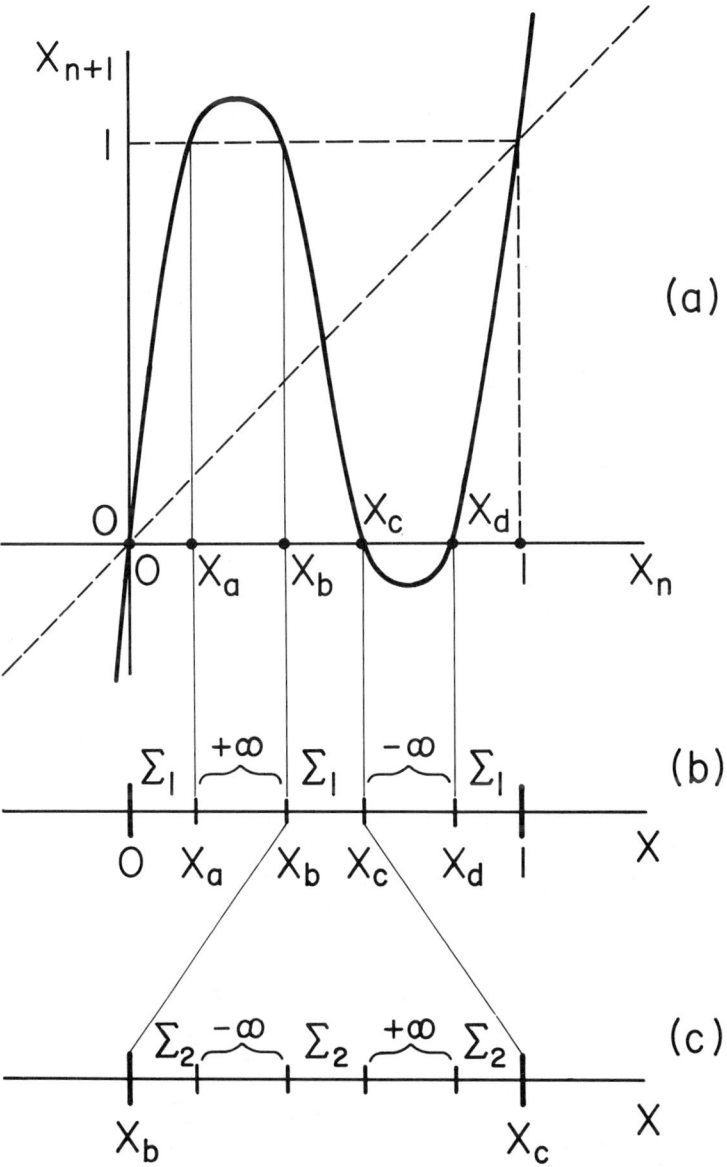

FIGURE 1. One-dimensional map.

2-D MAPS WITH UNIQUE BOUNDARY DIMENSION

Here, we consider a two-dimensional analogue of the situation in the previous section. We consider what is perhaps the simplest situation in which an invertible map in the plane can have two attractors with a fractal basin boundary separating their basins. Again, there is no self-similarity. While the reasons for the basin boundary to have fractal structure here are described in some detail in reference 2, the reasons for uniqueness of dimension are somewhat more subtle.

We will say f is a basin boundary horseshoe map if the action of the map f on a striplike region is as illustrated in FIGURE 2a. Specifically, we assume that f is a C^2 diffeomorphism (twice differentiable and having a differentiable inverse with continuous second-order partial derivatives). We assume that every point in region $\tilde{R}_1$ (and every point that is eventually mapped into $\tilde{R}_1$) has an orbit that is attracted to the attracting fixed point A_1. Similarly, orbits that enter $\tilde{R}_3$ are attracted to A_2. The image of region $\tilde{R}_2$ is an S-shaped strip. The left and right sides of $\tilde{R}_2$ map to the ends of the S. In addition, we assume only that the mapping is uniformly hyperbolic in $\tilde{R}_2$; that is, each point P, whose orbit remains in $\tilde{R}_2$ for all positive and negative times, has a contracting direction and an expanding direction, and the expanding (or contracting) direction of P is mapped by the Jacobian $df(x)/dx$ to the expanding (or contracting, respectively) direction of $f(P)$. The expansion is uniform in the sense that the expanding directions are expanded by the Jacobian by at least some factor $M > 1$. There is a similar uniformity of contraction on the contracting directions.

Theorem: The stable manifold of the invariant set of a uniformly hyperbolic basin boundary horseshoe map has a unique dimension.

Proof: The uniform hyperbolicity implies that there is a set C of points that remain in $\tilde{R}_2$ for all positive and negative times, and that each point of C is in the basin boundary. Also, each point of C has a stable manifold that is in the basin boundary. (Q is in the stable manifold of P if the distance between their n-th iterations goes to 0 exponentially fast as $n \to \infty$.) Each stable manifold stretches across the full height of $\tilde{R}_2$.

While the amount of stretching and contracting can vary somewhat from point to point in C, C resembles a product of Cantor sets. In fact, in $\tilde{R}_2$, the basin boundary B is a Cantor set of more or less vertical lines that are the stable manifolds of points of C. (If one follows the stable manifolds beyond the edges of $\tilde{R}_2$, almost all will bend around after leaving $\tilde{R}_2$ and will reenter it.) To gain further insight into the boundary structure and also to make the analogy with the previous section more clear, we consider a specially chosen region R_2 that lies within $\tilde{R}_2$. We construct R_2 by whittling away at the vertical edges of $\tilde{R}_2$. Due to the stretching, the ends of the S-shaped image of the whittled region will contract inward even faster than the whittled edges do, until the roughly vertical edges of the whittled region and those of its image coincide (FIGURE 2b). When this occurs, we stop whittling and denote the resulting region as R_2. The image of the left side of the reduced region R_2 is mapped into itself, and similarly for the right side. The process of reduction will typically result in nonvertical sides (cf. FIGURE 2b). Each of these sides contains a saddle-fixed point. In FIGURE 2c, we label curves to draw the analogy with FIGURE 1. There are two more or less vertical strips whose images under f are outside R_2. The left one (whose sides are labeled x_a and x_b) is mapped onto the right elbow of the S, and the right strip maps to the left elbow. The

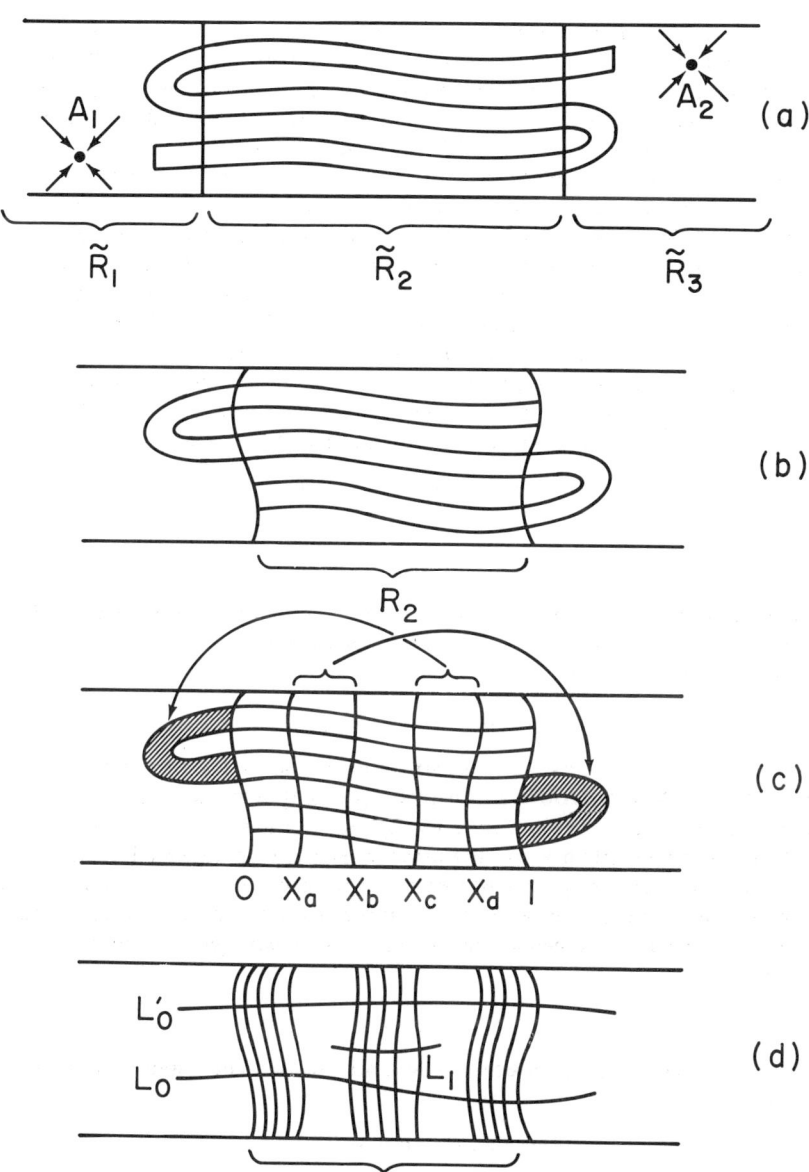

FIGURE 2. Two-dimensional map. (a) Action of the map on the rectangle $\tilde{R}_2$; A_1 and A_2 are two attractors. (b) Action of the map on a specially chosen reduced region R_2 that lies within $\tilde{R}_2$. (c) Regions between the curves through x_a and x_b and between the curves through x_c and x_d are in the basins of A_2 and A_1, respectively. (d) Basin boundary intersecting line segments L_0, L'_0, and L_1.

remaining three strips can be partitioned into nine strips running the full height of R_2, which remain in R_2 for two iterations. The set of points in the three strips will be called Σ_1 in analogy with the previous section, and the points in the nine strips are called Σ_2. Continuing the process, we see that the set of points that remain in R_2 for all positive time is the intersection of these collections of strips (that is, a Cantor set of curve segments), with each curve extending from the top edge of R_2 to its bottom.

In the previous section, the boundary B was a Cantor set. Here, we have an analogous Cantor set of curve segments. How this structure varies with height has been investigated by Newhouse.[6] Specifically, consider two curves, L_0 and L'_0, shown in FIGURE 2d. These curves cut all the way across R_2, crossing the stable manifolds transversally (that is, with nonzero angles), and they do so without doubling back. Newhouse[6] proves that there is a map h that carries L_0 onto L'_0 in such a way that if P on L_0 is on a stable manifold segment, then $h(P)$ lying on L'_0 is on the same stable manifold segment. He does this by showing that L_0 can be slid down to L'_0 with individual points of B sliding along their stable manifolds. Hence, by writing

$$B_0 = B \cap L_0 \quad \text{and} \quad B'_0 = B \cap L'_0,$$

we have

$$h(B_0) = B'_0.$$

He shows the mapping h is differentiable (with nonzero derivative). Letting δ_0 and δ'_0 denote the dimensions of B_0 and B'_0, respectively, it follows from property 2 (given earlier) that $\delta_0 = \delta'_0$ and

$$d(R_2) = 1 + \delta_0 = 1 + \delta'_0.$$

We now consider a curve L_k that extends across exactly one strip in one of the 3^k components of Σ_k, as in FIGURE 2d where $k = 1$. The k-th image of L_k will be a curve such as L_0 or L'_0 that stretches across R_2. We will call it L_0. The points B_k where L_k intersects B are mapped exactly onto B_0. Because the mapping is differentiable and invertible, $\delta_0 = \delta_k$, where δ_k is the dimension of B_k. Every neighborhood N of L_k that lies in R_2 will intersect B in a set of dimension $d(N)$ that is at least $1 + d(B_k)$. However, because the dimension cannot exceed the dimension of the set of all boundary points in R_2, we get that

$$d(N) = d(R_2).$$

Any open set U in R_2 that intersects B must contain some such N, and so the dimension of the boundary must satisfy

$$d(R_2) = d(N) = d(U).$$

We have argued here only for the case where U lies in R_2, but Newhouse's argument is also valid for the parts of the boundary that extend beyond R_2. Thus, it follows that the above equation is valid for any open set U that intersects B.

ASYMPTOTIC TRANSITIVITY ON THE BOUNDARY

The basin boundary structure in a region clearly depends on the long-term behavior of trajectories that start there. A principal idea of this paper is that if trajectories in different regions of the boundary have the same asymptotic behavior, then we may expect the dimension of the boundary to be the same in the different regions. If, however, the boundary points in one region have trajectories that stay away from the boundary trajectories of a second region, there is no reason why the boundary dimensions in those regions should be the same (unless the boundary in each region is a smooth surface). If the boundaries are fractal and if (by chance) the boundary dimensions of two such regions are the same, then we would expect to be able to make a small change in the system that changes the dimension of one, while having no influence on the trajectories through the other region. Thus, for typical systems, if the boundary trajectories from two regions stay away from each other and the boundary is fractal, the probability is zero that the boundary dimension in the two regions is the same.

Consider an initial condition x that is on the boundary of the basin of attraction and evolve it forward in time. Let $L^+(x)$ be the set of limit points generated by the forward time evolution of the boundary point x, and let Γ be the set of all limit points on the boundary (i.e., the union of all $L^+(x)$ for all x on the boundary). We call Γ the asymptotic set of the boundary, and we say that there is asymptotic transitivity on the boundary if there exists a trajectory that is dense in Γ; that is, if

$$L^+(x) = \Gamma$$

for some x.

Conjecture 1: The boundary has a unique dimension if the system is asymptotically transitive on the boundary. This conjecture is supported by our examples in the previous two sections.

For the 1-D map considered earlier, the basin boundary is a Cantor set, and it may be shown that there exist orbits dense in the boundary.[7] Thus, in this case, Γ is the whole boundary, and the boundary is asymptotically transitive. In conformity with our conjecture, it also has a unique dimension.

The basin boundary for the two-dimensional map in the previous section is generated by a standard horeshoe-type construction. In such a case, it is well known that there exists a set, λ, invariant under both forward and backward iterations of the map, that is the product of two Cantor sets and that has trajectories that are dense in this set.[8] From the construction of the previous section, the basin boundary is the stable manifold of this set and we may therefore identify λ with Γ. Thus, the basin boundary in the previous section has both asymptotic transitivity and a unique dimension, again in conformity with our conjecture.

We have discussed the transitive case, that is, where there is some x for which $L^+(x) = \Gamma$. In cases where the system is not asymptotically transitive in the boundary, we believe that typically, if Γ is bounded, there will be a finite number of "components" of Γ. We say Γ_i ($i = 1, 2, \ldots, N$) are the basic sets of Γ if they are disjoint, if their union is Γ, and if there exist points x_i such that

$$\Gamma_i = L^+(x_i), \quad i = 1, 2, \ldots, N.$$

In the case we discussed in conjecture 1, $N = 1$.

Furthermore, say that we have two points, x_a and x_b, for which $L^+(x_a)$ and $L^+(x_b)$ are contained in the same basic set Γ_i. Thus, we believe that if we take a sufficiently small neighborhood of x_a and a sufficiently small neighborhood of x_b, then the boundary dimension in these two neighborhoods is typically the same.

Conjecture 2: The number of possible dimension values for a basin boundary is at most the number N of basic sets Γ_i.

For axiom A dynamical systems (either flows or diffeomorphisms), the number of components of Γ will be finite.[9]

CONCLUSIONS

In this paper, we have considered the question of whether (and under what circumstances) the dimension of a basin boundary is unique. We have shown that there are common cases for which d is indeed unique. In particular, we have proven that there is a unique boundary dimension for one- and two-dimensional maps with horseshoe-type dynamics. Nevertheless, it is also true[4] that basin boundaries for other examples can have more than one dimension value. Thus, we have asked what type of boundaries have a single unique value of d. Based on our examples, we conjecture a condition for boundary dimension uniqueness—"asymptotic transitivity on the boundary" (see the previous section).

ACKNOWLEDGMENTS

We thank S. Newhouse and S-T. Yang for discussions.

REFERENCES

1. GREBOGI, C., S. W. MCDONALD, E. OTT & J. A. YORKE. 1983. Phys. Lett. **99A**: 415.
2. MCDONALD, S. W., C. GREBOGI, E. OTT & J. A. YORKE. 1985. Physica **17D**: 125.
3. MIRA, C. 1979. C. R. Acad. Sci. Paris **288A**: 591; GREBOGI, C., E. OTT & J. A. YORKE. 1983. Phys. Rev. Lett. **50**: 935; HOLT, R. G. & I. B. SCHWARTZ. 1984. Phys. Lett. **105A**: 327; SCHWARTZ, I. B. 1984. Phys. Lett. **106A**: 339; SCHWARTZ, I. B. 1985. J. Math. Biol. **21**: 347; TAKESUE, S. & K. KANEKO. 1984. Prog. Theor. Phys. **71**: 35; DECROLY, O. & A. GOLDBETER. 1984. Phys. Lett. **105A**: 259; MOON, F. C. & G-X. LI. 1985. Phys. Rev. Lett. **55**: 1439; GWINN, E. G. & R. M. WESTERVELT. 1985. Phys. Rev. Lett. **54**: 1613; GWINN, E. G. & R. M. WESTERVELT. 1986. Phys. Rev. **A33**: 4143; IANSITI, M., Q. HU, R. M. WESTERVELT & M. TINKHAM. 1985. Phys. Rev. Lett. **55**: 746; YAMAGUCHI, Y. & N. MISHIMA. 1985. Phys. Lett. **109A**: 196; NAPIORKOWSKI, M. 1985. Phys. Lett. **113A**: 111; ARECCHI, F. T., R. BADII & A. POLITI. 1985. Phys. Rev. **A32**: 402; MCDONALD, S. W., C. GREBOGI, E. OTT & J. A. YORKE. 1985. Phys. Lett. **107A**: 51; GREBOGI, C., E. OTT & J. A. YORKE. 1986. Phys. Rev. Lett. **56**: 1011.
4. GREBOGI, C., E. KOSTELICH, E. OTT & J. A. YORKE. Physica D. To be published.
5. PELIKAN, S. 1985. Trans. Am. Math. Soc. **292**: 695. This paper treats a one-dimensional map example and shows that $d_b = d_u$.
6. NEWHOUSE, S. 1979. Publ. Math. IHES **50**: 101.
7. For example, see: DEVANEY, R. L. 1986. An Introduction to Chaotic Dynamical Systems, p. 34–42. Benjamin. New York.
8. Ibid., p. 178–184.
9. GREBOGI, C., H. E. NUSSE, E. OTT & J. A. YORKE. Physica D. To be published.

Applications of the Semiclassical Spectral Method to Nuclear, Atomic, Molecular, and Polymeric Dynamics[a]

M. L. KOSZYKOWSKI,[b] G. A. PFEFFER,[c]
AND D. W. NOID[d]

[b]Combustion Research Facility
Sandia National Laboratories
Livermore, California 94550

[c]Department of Chemistry
University of Nebraska at Omaha
Omaha, Nebraska 68182-0109

[d]Chemistry Division
Oak Ridge National Laboratory
Oak Ridge, Tennessee 37831
and
Department of Chemistry
University of Tennessee
Knoxville, Tennessee 37996-1600

INTRODUCTION

Nonlinear dynamics plays a dominant role in a variety of important problems in chemical physics. Examples are unimolecular reactions,[1] infrared multiphoton decomposition of molecules,[2] the pumping process of the gamma ray laser,[3] dissociation of vibrationally excited state-selected van der Waals's complexes,[4] and many other chemical and atomic processes. The present article discusses recent theoretical studies on the quasi-periodic and chaotic dynamical aspects of vibrational-rotational states of atomic, nuclear, and molecular systems using the semiclassical spectral method (SSM).[5] We note that the coordinates, momenta, and so on, are found using classical mechanics in the studies included in this review. Consequently, certain processes of quantum mechanical origin, such as tunneling, have not been investigated in this framework. However, the projects to be described, undertaken using classical or semiclassical techniques, yield excellent results and explain many aspects of the underlying dynamics.

---

[a]This submitted manuscript has been authored by a contractor of the U.S. Government under Contract No. DE-AC05-84OR21400. Accordingly, the U.S. Government retains a nonexclusive, royalty-free license to publish or reproduce the published form of this contribution, or allow others to do so, for U.S. Government purposes. M. L. Koszykowski's work was supported by the U.S. Department of Energy. D. W. Noid's research was sponsored by the Division of Materials Sciences, Office of Basic Energy Sciences, U.S. Department of Energy, under Contract No. DE-AC05-84OR21400 with Martin Marietta Energy Systems, Inc.

The SSM technique can be applied to any autocorrelation function [e.g., of coordinates $x(t)$ or $y(t)$, momenta $p_x(t)$] or any dynamical variable [e.g., the dipole moment $\mu(t)$]. Below, we shall use $\chi(t)$ to denote any quantities that couple to a perturbation that causes a transition.

The infrared absorption-band shape function $I(\omega)$ [or in the case of any dynamical variable, $\chi(t)$, the power spectrum or spectral density] is related to the Fourier transform of its autocorrelation function $C(t)$ in the well-known way:

$$I(\omega) = \frac{1}{2\pi} \int_{-\infty}^{+\infty} C(t) \exp(-i\omega t)\, dt, \qquad (1.1)$$

where

$$C(t) = \langle \chi(0)\chi(t) \rangle. \qquad (1.2)$$

The average $\langle\ \rangle$ indicates an average over an ensemble appropriate to the problem (this ensemble will be described later). $C(t)$ has the stationary property

$$\langle \chi(0)\chi(t) \rangle = \langle \chi(\tau)\chi(t+\tau) \rangle, \qquad (1.3)$$

and because $\chi(t)$ is real, it also has the property of being an even function of t. For classical trajectories, equation 1.1 then reduces to

$$I(\omega) = \frac{1}{2\pi} \lim_{T\to\infty} \frac{1}{2T} \left\langle \left| \int_0^{2T} \chi(t) \exp(-i\omega t)\, dt \right|^2 \right\rangle \qquad (1.4)$$

in both the quasi-periodic and chaotic regimes described in the next paragraph. Equation 1.4 indicates that we can evaluate the spectra from a single classical trajectory generated from a time-independent Hamiltonian of the system studied. The choice of this trajectory will be discussed later.

The method has been most extensively used to characterize intramolecular dynamics. The understanding of nonlinear dynamical systems received tremendous attention during the late 1970s and the 1980s. During this time, it was found that a

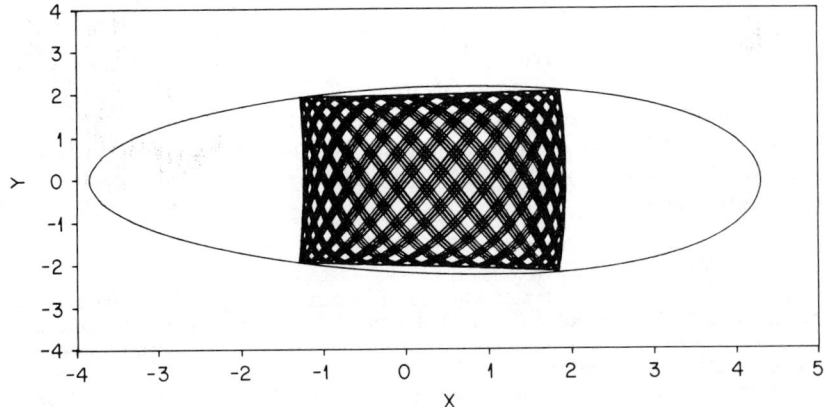

FIGURE 1. A typical boxlike quasi-periodic trajectory for a two-mode system.

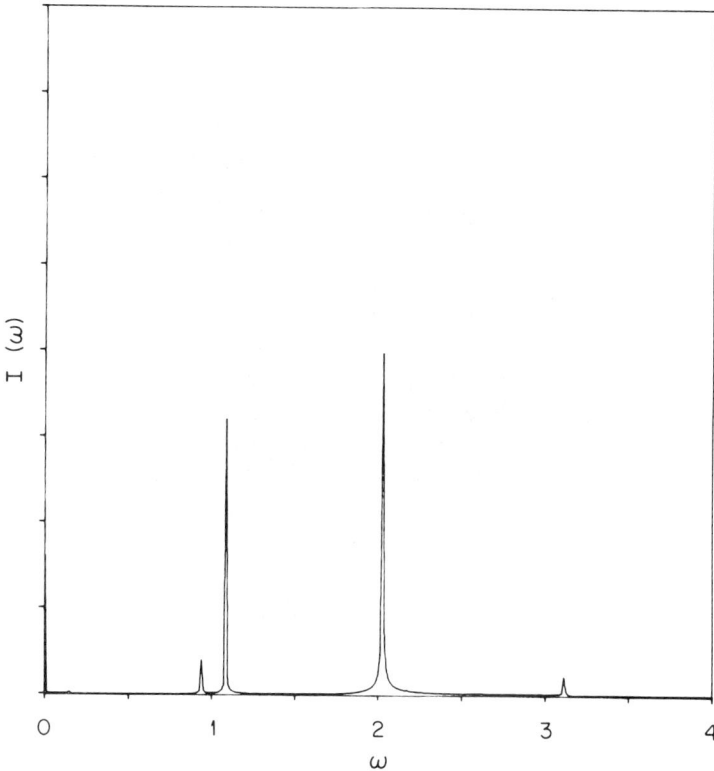

FIGURE 2. Spectra of the x and y coordinates of the trajectory shown in FIGURE 1.

dynamical system could exhibit two very different types of motion. A stable, more familiar type, termed quasi-periodic, is found for many ground-state systems. After a certain amount of excitation in the system, the motion becomes "random" and is termed chaotic. A recent review of molecular chaos by Noid, Koszykowski, and Marcus (NKM) can be found in reference 6. In the quasi-periodic regime, the motion can be decomposed into a Fourier series with a few fundamental frequencies and overtones. A typical quasi-periodic trajectory for a two-mode system is shown in FIGURE 1. The spectrum of the coordinates, say $x + y$, is shown in FIGURE 2 and, as expected, consists of a small number of sharp lines. In the chaotic regime, the trajectories sample much more (if not all) of the allowed phase-space, as shown in FIGURE 3. The spectrum, as expected, becomes broad and is shown in FIGURE 4. The use of correlation functions and spectra as tools for examining chaotic behavior has been extensively discussed in reference 6.

A number of possible choices have been explored for initial conditions of the trajectory used to evaluate equation 1.4. The methods are based upon quantized action integrals. Many years ago, Einstein[7] proposed a quantization of systems not permitting separation of variables: One finds canonical invariants, namely, the action variables J_i,

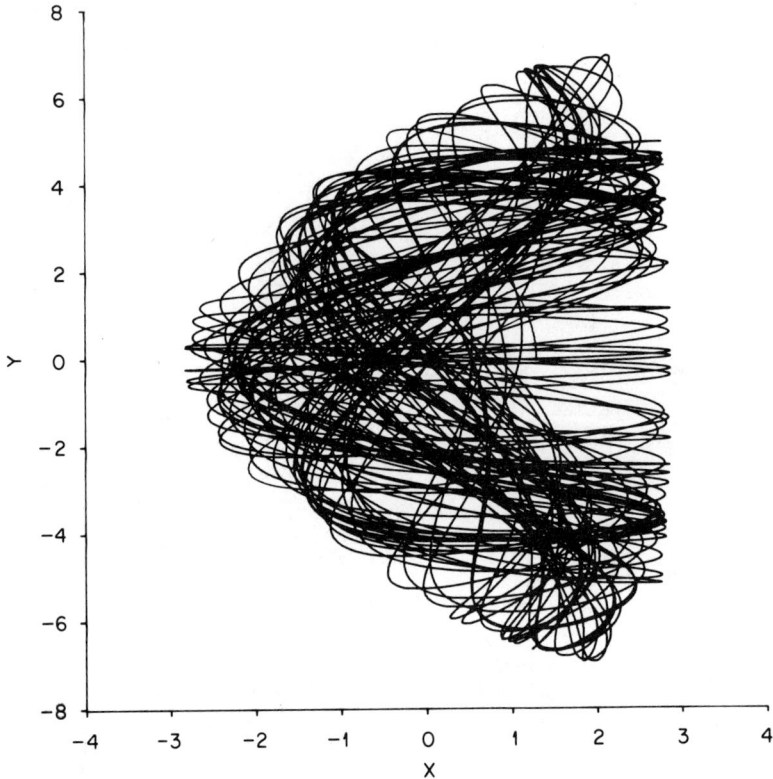

FIGURE 3. A chaotic trajectory for a two-mode system.

and quantizes them

$$J_i = \oint \tilde{p} \cdot d\tilde{q} = n_i h, \tag{1.5}$$

where the different J_i's are obtained by integrating over topologically independent paths; $\tilde{q}$, $\tilde{p}$ denote canonically conjugate coordinates and momenta, respectively. The theory was further developed by Keller,[8] who showed how fractional terms arose:

$$J_i = \oint \tilde{p} \cdot d\tilde{q} = (n_i + \delta_i)h, \tag{1.6}$$

where δ_i is a known constant, usually being 0 or $\frac{1}{2}$, depending on the degree of freedom. He further showed how to evaluate these integrals for a number of nonseparable systems that had zero potential energy within a confined region and that rose to infinity on the boundary of that region.

The SSM technique has been greatly extended in the past ten years, and now routine evaluation of equation 1.4 for multidimensional systems including resonances is possible. Originally, trajectories with $n_i = 0, 1, \ldots$ were used as initial conditions for the spectra from level n. Later, it was shown for the Morse oscillator that the use of

$\bar{n} = (n + m)/2$ for $n \to m$ transitions was preferable.[9] For other potential functions, slightly more accuracy can be obtained using other correspondence principles.

APPLICATION TO ATOMIC/MOLECULAR SYSTEMS

Polyatomic Molecules

The spectral analysis method has been applied to a large number of generic molecular Hamiltonians—the intent usually being to either develop or test new semiclassical techniques, or, more frequently, as a tool for studying the nonlinear dynamics and ramifications of chaos. A much smaller set of work has focused on Hamiltonians intended to represent a specific molecular or atomic system. In this section, we will discuss examples of both types of problems with an analysis on the work related most closely to molecular systems.

The choice of initial conditions for the trajectories is crucial to the success of this method. In their original work, NKM[5] used a primitive average of frequencies from zeroth order eigentrajectories. While this choice produced encouraging results, it was not until Koszykowski *et al.*[9] proposed that the $\bar{n} = (n_i + n_f)/2$ trajectory be used that a firm semiclassical basis for the initial condition was established. They applied this analytically to the Morse oscillator and demonstrated exact agreement with quantum calculations. In this work, it was also demonstrated, for the first time, that the spectral

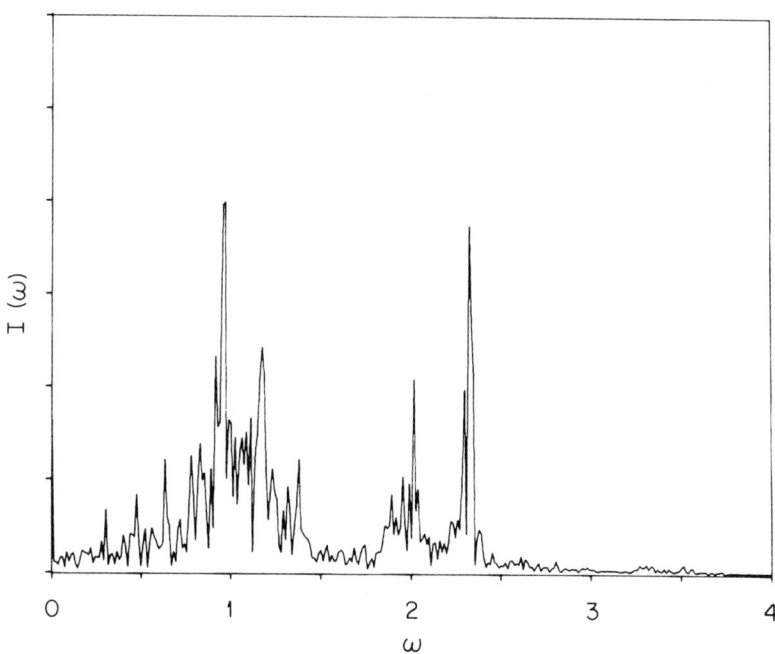

FIGURE 4. Spectra of the x and y coordinates of the trajectory shown in FIGURE 3.

analysis method could be used on a realistic surface where they calculated frequencies and intensities for OCS.[9] These workers also demonstrated that frequencies and intensities of overtones and combination bands could be easily obtained, even when they were many orders of magnitude weaker than the fundamental.[9]

Wardlaw et al.[10] have quantitatively tested the SSM technique on a wide variety of two-mode systems with several types of resonance interaction. The results were in very good agreement with exact quantum results and were considerably better than results obtained by low order perturbation theory. Later, Herbst and Noid[11] showed that even accurate bound-free and free-free transitions could be obtained for quasi-periodic–quasi-bound states.[12]

In another study, Liu, Noid, and Koszykowski[13] performed a spectral analysis of the vibrational motion of a realistic model of water. As the energy of the molecule increases beyond a certain transition value, the spectra changes from consisting of well-resolved lines to a much broadened one, indicating chaotic motion. Isotopic substitution that lowers the symmetry of the molecule reduced the transition energy because additional anharmonicities are introduced. Freezing the bending motion has a drastic effect on the motion of the molecules. Finally, a cross-correlation function was defined whose spectrum exhibits the same features as that of the self-correlation functions.

Stine and Noid[14] employed the spectral analysis method as a basis for studying diatomic molecules using numerically accurate potential functions. These potential functions were developed using the Rydberg-Klein-Rees (RKR) method for analyzing experimental spectral data. They calculated dipole matrix elements for HF, CO, and NO. They demonstrated that if the dipole function has a maximum at an intermolecular separation larger than the equilibrium configuration, the intensity of $\Delta n = 2$ transitions will be greater than $\Delta n = 1$ transitions for highly excited states. This is in complete contradiction to conventional wisdom based upon a harmonic oscillator model. They also presented an inversion method[15] where experimental results could be used to predict the dipole moment function. The method was applied to HF with excellent results.

A number of other gas-phase studies have applied the spectral analysis method to molecular problems. Hansel[16] related the width of broadened chaotic spectra to a value of the K entropy (see reference 6). He also discussed these concepts using a model of ozone. Smith et al.[17] also used the SSM technique on ozone and found the frequencies to be more accurate than an SSCF calculation. The semiclassical self-consistent field (SSCF) method was proposed a few years ago by Ratner and Gerber.[18] It reduces an n-coupled oscillator potential to n single-mode potentials that are determined self-consistently. These are then quantized semiclassically using the usual Bohr-Sommerfeld quantization rule. Later, Poppe[19] studied rotational spectra of SF_6 and also related them to the maximal Lyapunov exponent. Noid and Stine[20] and Martin and Wyatt[21] have all used this method in studies of IR multiphoton processes.

General studies of chaos using this method include Powell and Percival,[22] who computed a spectral entropy using the SSM technique. Farantos and Murrell,[23] Klemperer and co-workers,[24] Swamy and Hase,[25] Brickmann et al.,[26] Lopez et al.,[27] and Demontis et al.[28] all studied chaos in various molecular energy transfer processes. The study of chaos in molecular systems has been a much discussed topic, and the spectral analysis method proves to be a useful tool in characterizing different regions of phase-space.

In another study of vibrational spectroscopy, Koszykowski et al.[29] derived an expression for line shapes valid in both the quasi-periodic and chaotic regimes and for densities ranging from ultrahigh vacuum to several atmospheres. The Laplace transform of the dipole correlation function is given by

$$C(s) = C_0\left(s + \frac{i}{\tau}\right) \Big/ \left[1 - \left(\frac{1}{\tau}\right)C_0\left(s + \frac{i}{\theta}\right)\right], \qquad (2.1)$$

where C_0 is the zero pressure limit and τ is a relaxation rate due to collisions (calculated in the binary collision approximation). They found, as is well known, that for increasing collisional relaxation rates, the spectrum becomes broadened, shifted, overlapped, and eventually merged. A sequence of spectra for increasing collisional rates is shown in FIGURE 5 and clearly demonstrates this effect. Likewise, as the chaotic nature of the molecule increases, a similar effect is observed and the individual spectral lines merge more quickly. This can be clearly observed in FIGURE 6, as contrasted to FIGURE 5, where it is clear that chaos also affects the wings of the lines. Careful experiments at various pressures may be capable of distinguishing this effect from other complex variables in the line shape. Such a study would be an interesting way to observe chaos in the presence of collisions, along with being a generalization of the quantum beat experiments being done in a number of laboratories.

Koszykowski et al.[30] and Noid et al.[31,32] have also used correlation functions and spectra in new ways to extract fundamental information about the dynamics of molecular vibrations. In reference 30, they demonstrated that microcanonically averaged correlation functions decay from some initial value to that predicted by a microcanonical average over phase-space of the variables. Indeed, for these nonlinear systems, phase-space and time averages are interchangeable and the Rice-Ramsperger-Kassel-Marcus (RRKM) theory's basic assumption is valid.[1] RRKM is a theoretical treatment of unimolecular reactions. The first calculations of semiclassical intensities using primitive initial conditions are presented in reference 31, along with a derivation of intensities for the SSM technique. It was also demonstrated in this work that at a given total energy, one could locate chaotic spectra that are quite different. This led to the postulate that microcanonically averaged semiclassical spectra should correspond to quantum spectra generated from a wave packet composed of energies within a distance ΔE of the semiclassical state. Noid et al.[33] also have shown that the spectra of weak combination lines could be used to obtain the splitting between states degenerate in the primitive semiclassical method. In addition, by using correlation functions from classical trajectories, Muckerman et al.[34] demonstrated a method to determine the local-mode versus normal-mode character of molecular vibrations in a two-mode model of H_2O.

Very recently, we have introduced a method[33] for obtaining classical trajectories directly from a grid of points representing a potential energy function. Hamilton's equations were obtained by using an automated routine to generate multidimensional splines under tension through the total energy points. The output includes the derivatives at any arbitrary location in coordinate space that were used to numerically generate Hamilton's equations. The trajectories obtained with this method were indistinguishable from those generated with an analytic potential function and with analytic derivatives. This technique can now be used to couple directly with an electronic structure code to generate completely *ab initio* trajectories and spectra. No

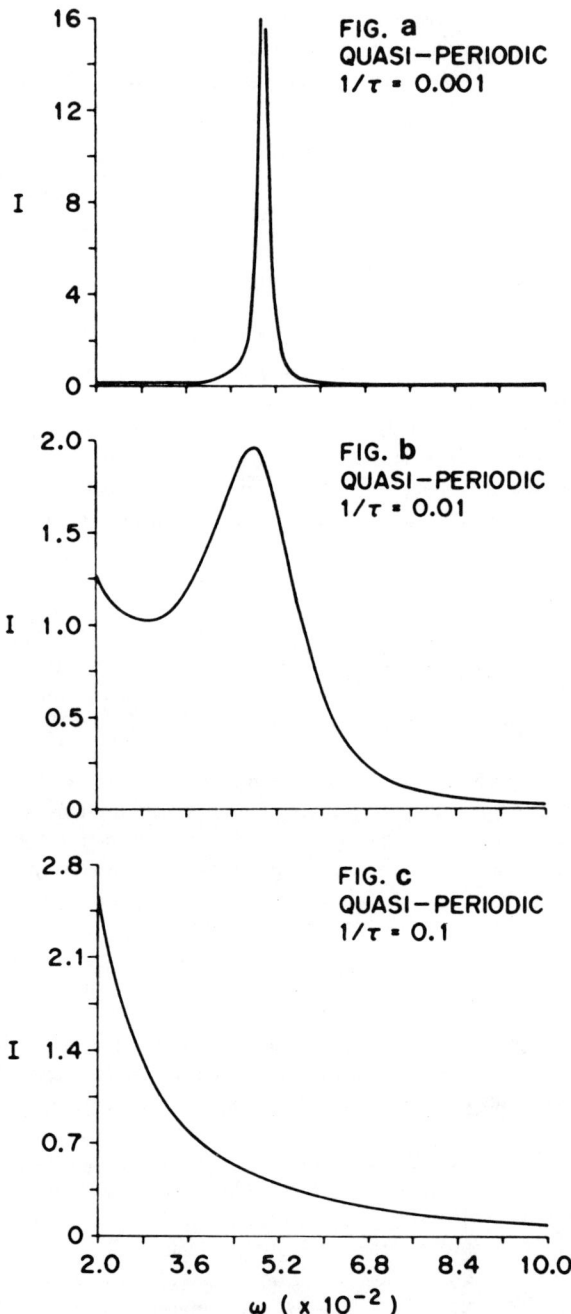

FIGURE 5. The dependence of the $\omega = 0.5$ peak for the Hénon-Heiles system with various values of collisional relaxation in the quasi-periodic regime.

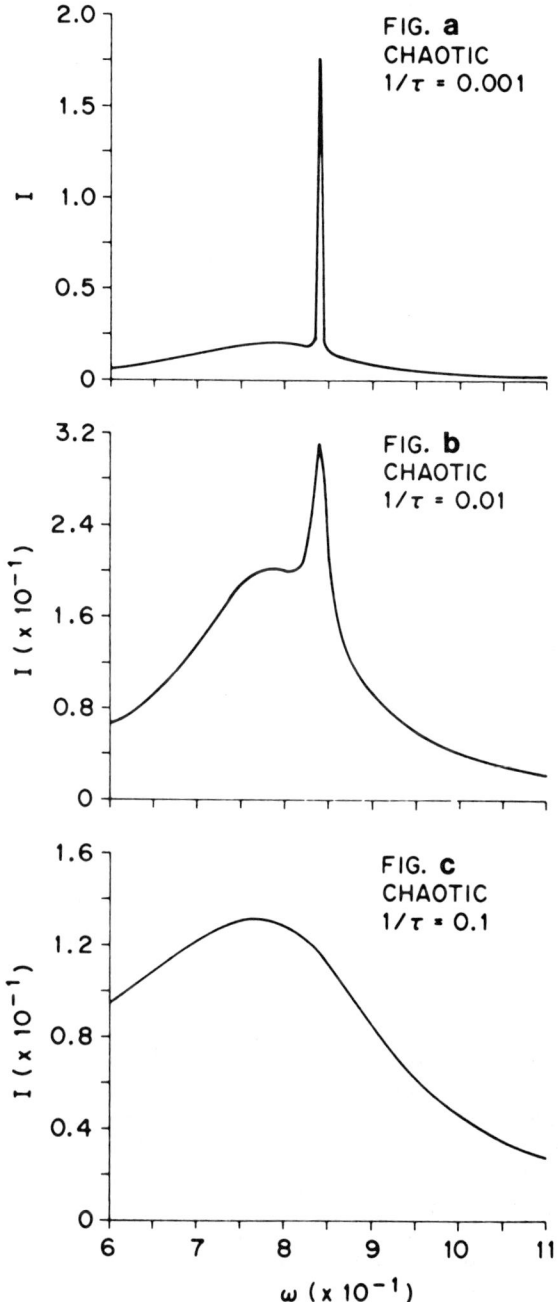

FIGURE 6. Same as FIGURE 5, but for chaotic motion.

analytic functions are imposed on the points and no errors are introduced by a fitting procedure. Thus, we can examine dynamics problems without the ambiguity of knowing a good representation of the potential function or what the best near separable coordinate system may look like, if there is one.

As a demonstration problem, we used fourth order many-body perturbation theory and a 6-31G* basis set[35] to calculate a grid of total energy points for ozone. This grid was combined with the spline under tension code to generate trajectories and spectra for ozone. A spectra for the ground state is shown in FIGURE 7. It is interesting to note that a harmonic analysis of the energy points predicts frequencies that are at least 100 cm^{-1} further from the experimental values than those in FIGURE 7. Analytic surfaces

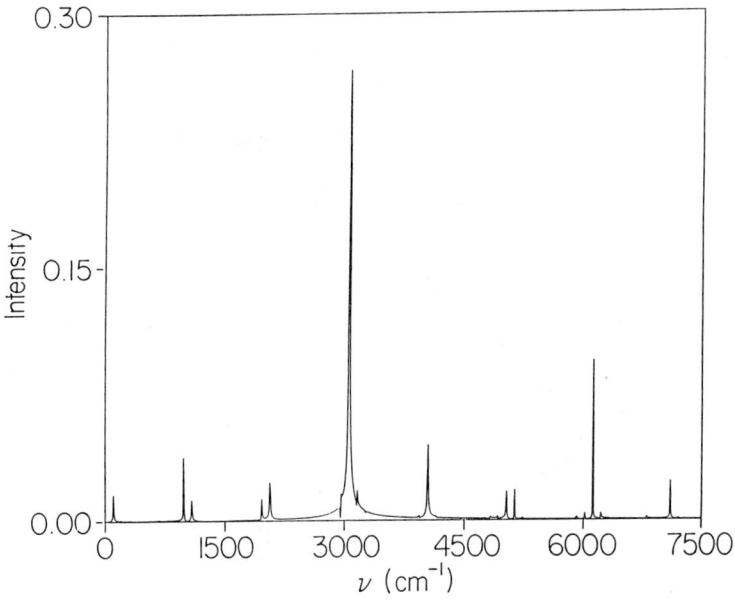

FIGURE 7. Spectra of a two-mode model of O_3 using a numerical grid of potential energy values and dipole moments for the ground state.

for ozone have shown significant regions of chaotic motion. In our study, we found no chaos at $E = 15,000$ cm^{-1}, which is the highest energy studied. Koszykowski et al.[36] applied the spectral method to examine the EELS spectra of ice multilayers on Ru. They used a polynomial potential fit to *ab initio* calculations of the local O-H mode energetics. The OH stretch frequency of hydrogen pointing into and out of the multilayer agreed with experiment to within 25 cm^{-1}. However, edge hydrogens in the calculation produced frequencies where there were clearly no experimental lines present. One possibility for the discrepancy is the limited potential used because electron structure calculations on metals are unreliable except in reproducing trends. Tully et al.[37] have recently extended the method to include a stochastic procedure aimed at vibrational relaxation. They performed a totally *ab initio* line-shape calculation for H on Si surfaces in excellent agreement with experiment. Gadzuk[38] and Adams[39] have both recently reported good results with the SSM technique for molecules absorbed on surfaces. Adams studied the vibrational dynamics of HCl on an Ar surface and calculated both the ro-vibrational and vibrational spectra. These later calculations were very successful in studying vibrational spectra on insular and semiconductor surfaces. A successful *ab initio* application to conductor surfaces must await further breakthroughs in the quality of electronic structure calculations on these systems.

Atomic and Nuclear Systems

The present method has also proven useful for calculation of atomic and nuclear spectra. Transition frequencies and intensities for excited states of atomic hydrogen in a strong magnetic field were calculated by the SSM technique.[40] The states considered had a principal quantum number of 30 and $L_z = 1$ for fields of up to 7 tesla. The spectra for states there that are labeled as librators and rotators are found to be qualitatively different, especially for the spectral component perpendicular to the field. In addition to the zero-field Kepler line, a new intramanifold transition was found at low frequency. The frequencies and intensities were found to be a sensitive function of the field strength and of the particular state of the Coulomb manifold involved.

A field-induced transition at low frequency was also observed, which is in agreement with predictions from perturbation theory; this agreement supports the accuracy of the computed intensities. The frequency and intensity are strongly dependent upon both the field strength and on the quantum state in the manifold. Qualitatively correct behavior is found for the intermanifold transition, but the accuracy of the calculation is not sufficient to identify the upper state uniquely. This method also found polarized spectra parallel and perpendicular to the field to be qualitatively different. In part, this effect is due, in quantum terms, to the z- and x-direction spectra obeying different selection rules. However, in addition, the distinction reflects the different natures of the orbits for the two classes of states mentioned above, which is demonstrated by the differences in the x-spectra. Spectral parameters in this transition regime would be difficult to obtain by other methods.

Finally, in this subsection, we now discuss a coupled nucleon electronic system. In order to model the energy transfer from an inner core electron to an excited nuclear proton, we have chosen to use the nuclear shell model approach for an independent valence nucleon. This model serves to introduce the important characteristics—primarily energy level spacing and nuclear size effects—which are orders of magnitude disparate from the electronic characteristics. In this model, a valence excited proton is bound as an independent particle to the nuclear core using an effective Wood Saxons potential.[41] The electron is bound to both the core of protons and the valence proton using the normal Coulomb potential. Our Hamiltonian is

$$H = \frac{1}{2m_N}\left(P_N^2 + \frac{L_N^2}{R_N^2}\right) + V_N(R_N) + \frac{1}{2m_e}\left(P_e^2 + \frac{L_e^2}{r_e^2}\right) - \frac{Z}{r_e^2} + H_{\text{INT}}, \quad (2.2)$$

where m_N is the mass of a nucleon, P_N is the nucleon momentum, L_N is the nucleon orbital angular momentum, R_N is the radial position of the nucleon, and

$$V_N(R_N) = V_0/(1 + e^{(R_N-R)/A}) \tag{2.3}$$

is the nucleon potential. In this potential, V_0 is the well depth and A is the surface diffusivity parameter. The quantities, m_e, P_e, L_e, and r_e, are the corresponding electron properties. The Coulomb potential is included for the interaction of the electron with the closed shell nuclear core and is given by

$$H_{\text{INT}} = -\frac{1}{r_{ep}}, \tag{2.4}$$

where $r_{ep} = |\tilde{r}_e - \tilde{R}_N|$.

In our study, we considered the coupling of the 1S electronic state with the excited nucleon state for closed shell +1 nucleons. The number of protons considered are 29, 51, and 83. In FIGURE 8, we present the unpolarized spectra for the coupled electron-nucleon model, with the electronic state being 1S and a 3F proton state with $Z = 83$. In FIGURE 9, we present the uncoupled spectrum for the same nuclear-electron state.

As can be observed from these figures, a large number of lines with spacing corresponding to electronic transition are observed. A smaller number of larger lines occur at about 1 and 3 MeV (not shown) that correspond to pure nuclear transitions. Interestingly enough, the spectra in FIGURE 8 appear to have some features of chaos,

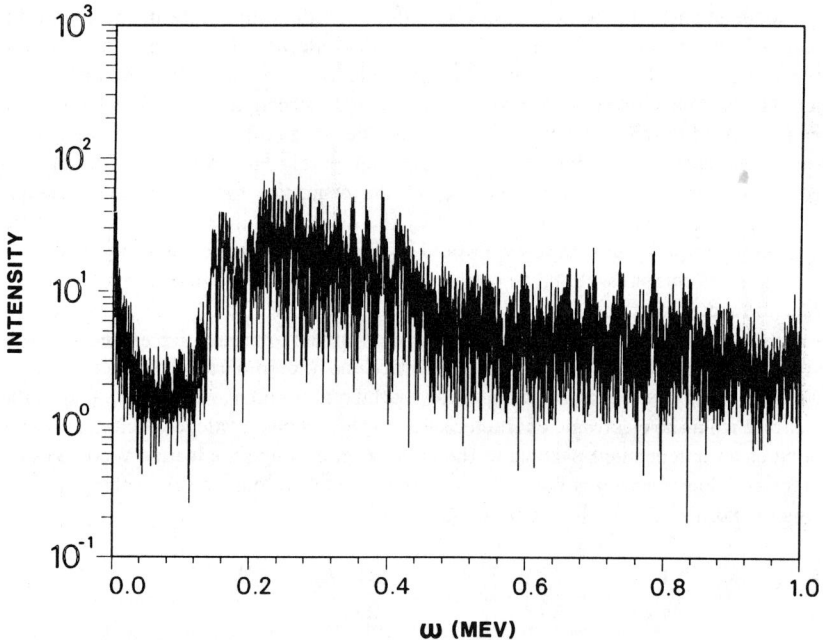

FIGURE 8. Unpolarized spectra for the nuclear-electron model showing chaotic motion.

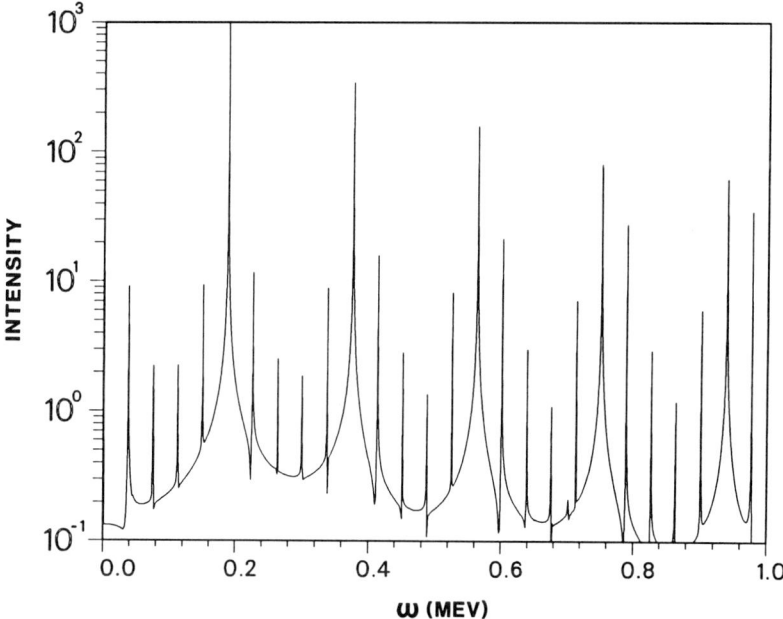

FIGURE 9. Same as FIGURE 8, but for the uncoupled case.

which indicates that the classical dynamics is strongly coupled. For the $Z = 29$ case, none of the spectra appear to have chaotic features, and therefore the coupling is much weaker.

Among the preliminary conclusions of this study are:[42]

— high Z produces the greatest distortion of the nuclear-electron spectra;
— use of the Langer correction[43] for the 1S state is very important (because old quantum theory orbits and spectra failed to be chaotic);
— finally, the SSM technique is a very practical technique to study the very subtle coupling in this model.

Polymeric Systems

In this section, we present some preliminary results concerning the transition from quasi-periodic to chaotic motion in a long linear diatomic chain. A few studies have been reported for linear atomic chains,[44] but even less work has been reported using polymeric diatomic chains.[45] We also further extended the previous research by studying longer chain lengths: the longest chains considered before had contained ~1000 atoms; we have employed 20–3200 atoms (10–1600 diatoms). The motivation for both our investigations and those preceding us has been twofold: to find an appropriate criteria to detect or predict the onset of chaos in systems with many modes, and to examine the stochastic threshold as a function of chain length, initial excitation, and other chain attributes. Such work is applicable to energy transfer in polymers and in solids.

Our model chain mimics a linear array of AB molecules where the A—B bond is a typical chemical bond and the $B \cdots A$ interaction is much weaker. A Morse oscillator is chosen for the AB bond and has the form,

$$V(r) = D(e^{-2\alpha(r-re)} - 2e^{-\alpha(r-re)}), \qquad (2.5)$$

where r is the A—B distance. The $B \cdots A$ interaction is given by a Lennard-Jones potential,

$$V(R) = 4\epsilon\left[\left(\frac{\sigma}{R}\right)^{12} - \left(\frac{\sigma}{R}\right)^{6}\right], \qquad (2.6)$$

where R is the distance between the molecular centers of mass. Our choice of parameters is appropriate for a chain of HF molecules. In the HF chain, the $F \cdots H$ interaction is an example of hydrogen bonding and the center of mass of the molecule can be considered to be at the F nucleus.

A one-dimensional "temperature" was calculated from the kinetic energy at each time step. This temperature remained reasonably constant after an initial equilibration period. The spectra were generated by a fast Fourier transform of the sum of all the particle coordinates at each time step. The initial H and F positions were first chosen so that each particle sat at the minimum distance of the Morse oscillator and the Lennard-Jones potential relative to that particle's nearest neighbor. A random displacement (either positive or negative) from that minimum position was then

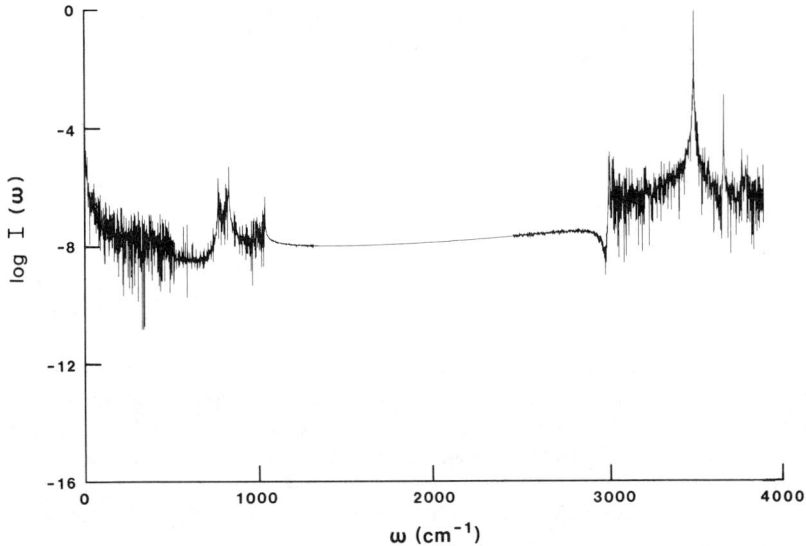

FIGURE 10. Spectra of a polymeric HF chain with 1280 atoms in the quasi-periodic regime for a low temperature (1 K).

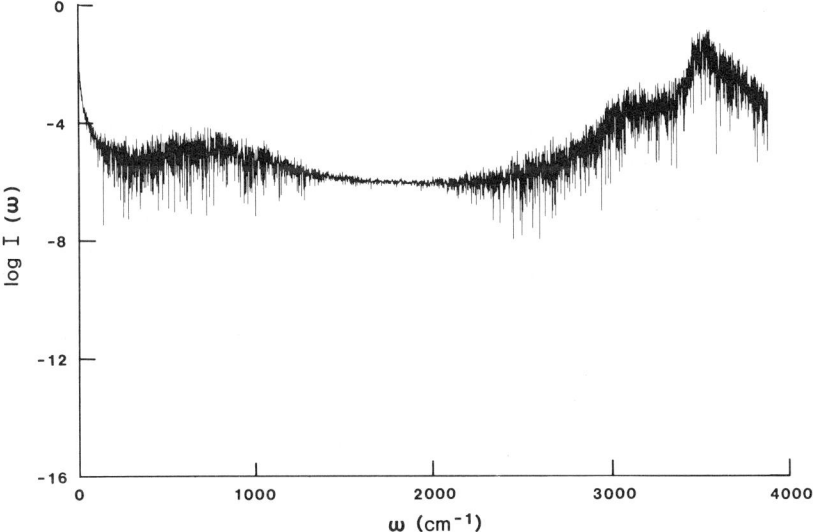

FIGURE 11. Same as FIGURE 10, but for a higher temperature (135 K) and chaotic motion.

introduced to give the final positions. By changing the maximum position displacement parameter and the maximum momentum parameter, the temperature could be changed. The ends of the chain were not fixed (in contrast to several previous studies).

The spectra clearly showed a low frequency set of peaks ($\lesssim 100$ cm^{-1}) and a high frequency set of peaks (~ 3000 cm^{-1}). The choice of initial conditions described above gives, in the low temperature case, nearly 100% excitation of the high frequency modes. The conclusions reported here deal only with these initial conditions, and conclusions for other choices of initial conditions may be different.

FIGURE 10 shows the power spectrum for 1280 particles (640 HF molecules) at a nominal temperature of 1 K. The log of the intensity from the fast Fourier transform is actually displayed rather than the intensity itself because the low frequency peaks are more easily seen in the log plot. Well-separated sharp peaks can be seen in the range 0–1300 cm^{-1} and in the range 3250–3924 cm^{-1}. There is an intermediate region where little noise and no peaks are seen. Based upon the appearance of the spectrum, the system appears to be quasi-periodic at this temperature. FIGURE 11 shows the (log) power spectrum for the same system at 135 K, where no sharp peaks are seen and where the noise-free intermediate region has disappeared. Based upon the appearance of this spectrum, the system appears to be chaotic.

Furthermore, if the integral of the spectrum is taken from 0–$\sim$1982 cm^{-1} (the low frequency regime) and a second integral is taken from 1982–3924 cm^{-1} (the high frequency regime), then for FIGURE 10, the low frequency/high frequency percentages are 0/100, whereas for FIGURE 11, the ratio is 24/76. Clearly, the initial excitation (which evidently was of the high frequency modes only because the low temperature spectrum shows only high frequency peaks) has remained localized in the quasi-periodic case and has begun to redistribute among all modes in the chaotic case. In solid polymeric materials, the low frequency modes are known as acoustical modes and the high frequency modes are known as optical modes. The technique proposed here should be useful in studying the onset of efficient energy transfer from the optical modes (perhaps excited by laser light) to acoustical modes (which can lead, upon excitation, to damage in the solids).

We note, as a final comment, that in our studies we have found that the temperature at which the chaotic limit is reached has a more complicated dependence upon the chain length than has been seen before for the atomic chains. Preliminary calculations indicate that the temperature at which the spectra become chaotic increases with increasing chain length up to a point and then essentially remains constant or drops somewhat as the chain length is further increased. Work is in progress to explore these results more fully, as well as applying these energy transfer ideas to solid polyethylene and graphite.

CONCLUSIONS

In this discussion, we have outlined the semiclassical spectral method and a wide variety of applications. Although this technique was first developed ten years ago, it has proved to be tremendously successful as a tool used in dynamics problems. Applications include problems in nonlinear dynamics, molecular and atomic spectra, surface science, astronomy and stellar dynamics, nuclear physics, and polymer physics. Innovation in the high resolution spectral method in place of the fast Fourier transform may, in the future, make possible more accurate calculations with shorter trajectories. As utilization of this technique becomes more routine, we expect many other applications to be reported in the scientific literature.

ACKNOWLEDGMENTS

We wish to acknowledge R. A. Marcus, W-K. Liu, S. K. Knudson, J. D. McDonald, J. R. Stine, and F. X. Hartman for numerous helpful discussions. We also thank Ceci Steele for her excellent secretarial help.

REFERENCES

1. ROBINSON, P. J. & K. A. HOLBROOK. 1972. Unimolecular Reactions. Wiley. New York; FORST, W. 1973. Unimolecular Reactions. Academic Press. New York.
2. AMBARTZUMIAN, R. V. & V. S. LETOKHOV. 1977. Chemical and Biochemical Applications of Lasers. Vol. 3. C. B. Moore, Ed.: 1. Academic Press. New York; LETOKHOV, V. S. 1977. Ann. Rev. Phys. Chem. **28**: 133; SCHULZ, P. A., AA. S. SUDBØ, D. J. KRAJNOVICH, H. S. KWOK, Y. R. SHEN & Y. T. LEE. 1979. Ann. Rev. Phys. Chem. **30**: 379.
3. NOID, D. W., F. X. HARTMANN & M. L. KOSZYKOWSKI. 1986. In Advances in Laser Science II. W. C. Stwalley, Ed. AIP Conf. Proceedings.
4. LEVY, D. H. 1980. Ann. Rev. Phys. Chem. **31**: 197.
5. NOID, D. W., M. L. KOSZYKOWSKI & R. A. MARCUS. 1977. J. Chem. Phys. **67**: 404–408.
6. NOID, D. W., M. L. KOSZYKOWSKI & R. A. MARCUS. 1981. Ann. Rev. Phys. Chem. **32**: 267.

7. EINSTEIN, A. 1917. Verh. Dtsch. Phys. Ges. **19:** 82.
8. KELLER, J. B. 1958. Ann. Phys. N.Y. **4:** 180.
9. KOSZYKOWSKI, M. L., D. W. NOID & R. A. MARCUS. 1982. J. Phys. Chem. **86:** 2113–2117.
10. WARDLAW, D., D. W. NOID & R. A. MARCUS. 1984. J. Phys. Chem. **88:** 536–547.
11. HERBST, E. & D. W. NOID. 1984. Chem. Phys. Lett. **109:** 559–562.
12. NOID, D. W. & M. L. KOSZYKOWSKI. 1980. Chem. Phys. Lett. **78:** 114–117; NOID, D. W., S. K. KNUDSON & J. B. DELOS. 1983. Chem. Phys. Lett. **100:** 367–370.
13. LIU, W-K., D. W. NOID & M. L. KOSZYKOWSKI. 1982. *In* Intramolecular Dynamics. J. Jortner & B. Pullman, Eds.: 191–203. Holland. Dordrecht.
14. STINE, J. R. & D. W. NOID. 1983. J. Chem. Phys. **78:** 1876–1883.
15. STINE, J. R. & D. W. NOID. 1983. J. Chem. Phys. **78:** 3647–3651.
16. HANSEL, K. D. 1978. Chem. Phys. **33:** 35; HANSEL, K. D. 1979. Laser Induced Processes in Molecules. K. L. Kompa & S. D. Smith, Eds.: 145. Springer-Verlag. Berlin/New York.
17. SMITH, A. D., W-K. LIU & D. W. NOID. 1984. Chem. Phys. **89:** 345–352.
18. RATNER, M. A. & R. B. GERBER. 1979. Chem. Phys. Lett. **68:** 195.
19. POPPE, D. 1980. Chem. Phys. **45:** 371.
20. NOID, D. W. & J. R. STINE. 1982. J. Chem. Phys. **76:** 4947–4951.
21. MARTIN, D. L. & R. E. WYATT. 1982. Chem. Phys. **64:** 203.
22. POWELL, G. E. & I. C. PERCIVAL. 1979. J. Phys. **A12:** 2053.
23. FARANTOS, S. C. & J. N. MURRELL. 1981. Chem. Phys. **55:** 205.
24. LEHMANN, K. K., G. J. SCHERER & W. KLEMPERER. 1982. J. Chem. Phys. **76:** 6441.
25. SWAMY, K. N. & W. L. HASE. 1982. Chem. Phys. Lett. **92:** 371.
26. BRICKMANN, J., R. PFEIFFER & P. C. SCHMIDT. 1984. Ber. Bunsenges. Phys. Chem. **88:** 382.
27. LOPEZ, V., V. FAIREN, S. M. LEDERMAN & R. A. MARCUS. 1986. J. Chem. Phys. **84:** 5494.
28. DEMONTIS, P., G. B. SUFFRITTI, E. S. FOIS & A. GAMBA. 1986. Chem. Phys. Lett. **127**(N5): 456–461.
29. KOSZYKOWSKI, M. L., W-K. LIU & D. W. NOID. 1982. J. Chem. Phys. **77:** 2836–2840.
30. KOSZYKOWSKI, M. L., D. W. NOID, M. TABOR & R. A. MARCUS. 1981. J. Chem. Phys. **74:** 2530–2535.
31. NOID, D. W., M. L. KOSZYKOWSKI, M. TABOR & R. A. MARCUS. 1980. J. Chem. Phys. **72:** 6169–6175.
32. NOID, D. W., M. L. KOSZYKOWSKI & R. A. MARCUS. 1983. J. Chem. Phys. **78:** 4018–4024.
33. NOID, D. W., S. K. KNUDSON, M. L. KOSZYKOWSKI & R. J. RENKA. 1986. J. Phys. Chem. **90:** 6135–6138.
34. MUCKERMAN, J. T., D. W. NOID & M. S. CHILD. 1983. J. Chem. Phys. **78:** 3981–3989.
35. The latest version of the Gaussian 80 code is available from QCPE.
36. KOSZYKOWSKI, M. L., P. THIELE, F. HOFFMAN, J. S. BINKLEY & D. W. NOID. 1982. J. Am. Vac. Soc.
37. TULLY, J. C., Y. J. CHABAL, K. RAGHAVACHARI, J. M. BOWMAN & R. R. LUCCHESE. 1985. Phys. Rev. B. Condensed Matter **31:** 1184.
38. GADZUK, J. W. 1986. J. Electron Spectrosc. Relat. Phenom. **38:** 233.
39. ADAMS, J. E. 1986. J. Chem. Phys. **84:** 3589.
40. KNUDSON, S. K. & D. W. NOID. 1984. Chem. Phys. **89:** 353–361.
41. DAVYDOV, A. S. 1968. Quantum Mechanics, section 94. Addison–Wesley. Reading, Massachusetts.
42. KOSZYKOWSKI, M. L., D. W. NOID, F. X. HARTMANN & G. A. PFEFFER. 1987. J. Nucl. Phys. In press.
43. LANGER, R. G. 1937. Phys. Rev. **51:** 669.
44. DIANA, E., L. GALGANI, M. CASARTELLI, G. CASATI & A. SCOTTI. 1977. Theor. Math. Phys. **29:** 1022; CAROTTA, M. C., C. FERRARIO, G. LOVECCHIO & L. GALGANI. 1978. Phys. Rev. **A17:** 786; JACKSON, A. 1978. Rocky Mount. J. Math. **8:** 127; GALGANI, L. & G. LOVECCHIO. 1979. Nuovo Cimento **B52:** 1.
45. HENRY, B. I. & J. OITMAA. 1985. Aust. J. Phys. **38:** 171; HENRY, B. I. & J. OITMAA. 1985. Aust. J. Phys. **38:** 191.

Kelvin-Helmholtz Instabilities in the Interstellar Medium

JAMES H. HUNTER, JR.[a] AND RODNEY W. WHITAKER[b]

[a]*Department of Astronomy*
University of Florida
Gainesville, Florida 32611

[b]*Earth and Space Science Division*
Los Alamos National Laboratory
Los Alamos, New Mexico 87545

INTRODUCTION

Our present understanding of the interstellar medium (hereafter designated ISM) reveals that the gas is turbulent and that several phases of the ISM may coexist, spanning large ranges of temperature and density. Relatively cool HI regions have number densities $n > 1$ cm^{-3} and temperatures $T < 10^2$ K,[1] whereas the corresponding figures in molecular clouds are $n \gtrsim 300$ cm^{-3} and $T <$ a few $\times$ 10^2 K;[2] in dense molecular clouds, $T < 10$ K.[2]

While the detailed velocity structure of the ISM has been mapped only selectively, an appropriate description of the medium might be "a continuous distribution of supersonic gas streams and turbulent eddies".[3] The observed gas velocities range from small to $\sim 10^3$ km s^{-1}, with characteristic turbulent speeds[4] of the order of 10 km s^{-1}. Consequently, the Mach numbers, M, in the ISM span an enormous range: $0 < M < 10^4$, say. Notwithstanding, a more typical range might be $0 < M < 50$.

The physical state of the ISM is so different from that of our direct experience that familiar treatments of terrestrial and laboratory phenomena, such as Rayleigh-Taylor and Kelvin-Helmholtz instabilities, must be examined critically before applying them to the interstellar environment. An appreciation of the importance of several physical processes may be gained by examining their typical characteristic time scales in the ISM. In order of increasing importance, three important time scales are:

(1) The sound travel time, $t_s = \lambda/c$, where λ is the wavelength of a disturbance and c is the local sound speed.
(2) The dynamical time scale, $t_d = \lambda/U$, where U is the bulk flow speed of the gas. Hence, unless $M = 0$, $t_d = t_s/M$. In the limiting case of an incompressible fluid, $M = 0$. For compressible fluids, the Mach number may be thought of as a gauge of the relative importance of compressibility. While the incompressible approximation is useful in many terrestrial applications in which the flow speeds are small, clearly it cannot be valid in the ISM where typically $M > 1$.
(3) The characteristic cooling time, $t_c = \mathcal{U}/\mathcal{C}$, where $\mathcal{U}$ and $\mathcal{C}$ are, respectively, the internal energy and cooling rate of the gas. In relatively dense, optically thin regions of the ISM, t_c may be very much shorter than t_s.

Consequently, on the time scales available for dynamical changes to occur in the ISM, the fluid behavior cannot be remotely approximated as adiabatic. In the following discussion, we will show, in treating the instability problem, that it is not possible to use an isothermal equation of state either, no matter how intuitively appealing that notion may seem. In the present paper, we will concentrate upon the implications of the above three time scales. Other important time scales that must be considered in more generalized and realistic treatments of the ISM include the free-fall time scale, the time scale for ambipolar diffusion, the characteristic time scale of shear viscous transport, and the local period of galactic rotation (along with the local galactic shearing rate).

The crucial role of cooling in the growth of instabilities in the ISM has been demonstrated by Hunter et al.,[3] who studied the head-on collisions of molecular gas streams. Rayleigh-Taylor–like instabilities developed at the interfaces of the colliding flows when molecular cooling was included in the calculations, whereas the interface was stable when the gas was not allowed to cool. The strong molecular cooling lowered the pressures at the interfaces sufficiently to destabilize them. These instabilities grew dramatically in the nonlinear regime; the ultimate source of the driving energy was the kinetic energy of the streams. Inasmuch as gas flows in the ISM would seldom collide head-on, it seems inevitable that Kelvin-Helmholtz–like instabilities (or, more generally, hybrid, Kelvin-Helmholtz, Rayleigh-Taylor instabilities) would arise near their interfaces due to the presence of velocity shear. Moreover, in light of the work of Hunter et al.,[3] it seems probable that the thermodynamic effects of strong cooling are of great importance in these instabilities.

The present work is a first step in this direction. Within the framework of classical, linear analysis, we examine the growth of Kelvin-Helmholtz–like (hereafter denoted K-H–like) instabilities when simple, local heating and cooling mechanisms operate in the media. Thus, our work includes the physics of both classical thermal[5,6] and compressible K-H instabilities.[7–9] We will show that the resultant instabilities have novel and unexpected characteristics.

THE LINEAR ANALYSIS

As the unperturbed condition, we adopt the simplest possible equilibrium state, which consists of a velocity discontinuity, or vortex sheet, with constant, but generally unequal, gas densities and temperatures on either side of the discontinuity. While this model has some limitations, especially for compressible fluids,[10] it has the virtue of allowing exact solutions in three limiting cases of interest. Discontinuities in both density and temperature should occur naturally in the ISM. As first shown by Field,[5] thermal instabilities can take place in the medium, which will result in two thermally stable phases coexisting in pressure equilibrium. One component is of relatively low temperature and high density, whereas the other is of relatively high temperature and low density. Each phase is in thermal equilibrium—meaning that the local heating and cooling rates balance. In HI regions, thermal conduction is so inefficient that the boundaries between the two phases may be regarded as discontinuous.

We define an inertial, Cartesian coordinate system to be at rest with respect to the fluid occupying the lower-half space (extending from $z = 0$ to $z = -\infty$). The problem is depicted in FIGURE 1. We adopt the following notation:

ρ = bulk gas density (g cm^{-3}),
T = gas kinetic temperature (K),
p = gas pressure (erg cm^{-3}),
u, v, w = velocity components in the x, y, and z directions, respectively (cm s^{-1}),
U = constant magnitude of x velocity in medium 1,
γ = the ratio of the specific heat at constant pressure to that at constant volume,
$c = \sqrt{\gamma p/\rho}$, the adiabatic sound speed (cm s^{-1}),
$\mathcal{H} = \mathcal{H}(\rho, T)$, the local gas heating rate (erg cm^{-3} s^{-1}),
$\mathcal{C} = \mathcal{C}(\rho, T)$, the local gas cooling rate (erg cm^{-3} s^{-1}),
$k_{x,y} = 2\pi/\lambda_{x,y}$, the perturbation wave numbers in the x and y directions (cm^{-1}),
$k = \sqrt{k_x^2 + k_y^2}$, the total horizontal wave number,
$\theta = \cos^{-1}(k_x/k)$, the angle between k and the x-axis,
$M = U/c_1$, the Mach number characterizing the flow in region 1, and
$m = M \cos \theta$, the effective Mach number.

For the equilibrium temperatures to be maintained, we require that $\mathcal{H} = \mathcal{C}$ in each medium.

Our analysis parallels the logic outlined by Gerwin[7] and generalized by Hunter.[9] We disturb the fluids, letting the density, pressure, and velocity components be $\rho + \rho'$, $p + p'$, $U + u'$, v', and w', respectively, where the unadorned quantities are the constant equilibrium values and the primed quantities are small Eulerian perturbations. In each medium, the linearized equations of continuity, motion, and thermal energy balance are

$$\mathcal{D}\rho' = -\rho \left(\frac{\partial u'}{\partial x} + \frac{\partial v'}{\partial y} + \frac{\partial w'}{\partial z} \right), \tag{1}$$

$$\rho \mathcal{D} u' = -\frac{\partial p'}{\partial x}, \tag{2}$$

$$\rho \mathcal{D} v' = -\frac{\partial p'}{\partial y}, \tag{3}$$

$$\rho \mathcal{D} w' = -\frac{\partial p'}{\partial z}, \tag{4}$$

and

$$(\mathcal{D} + \mathcal{F}_p) p' = \frac{\gamma p}{\rho} (\mathcal{D} + \mathcal{F}_\rho) \rho', \tag{5}$$

where the operator

$$\mathcal{D} \equiv \frac{\partial}{\partial t} + U \frac{\partial}{\partial x}.$$

The thermodynamic quantities appearing in the energy equation are defined by

$$\mathcal{F}_p \equiv (\gamma - 1)\left(\frac{\mathcal{C}}{p}\right)\left(\left.\frac{\partial \ln \mathcal{C}}{\partial \ln T}\right|_\rho - \left.\frac{\partial \ln \mathcal{H}}{\partial \ln T}\right|_\rho\right) \tag{6}$$

and

$$\mathcal{F}_\rho \equiv \left(\frac{\gamma - 1}{\gamma}\right)\left(\frac{\mathcal{C}}{p}\right)\left(\left.\frac{\partial \ln \mathcal{C}}{\partial \ln T}\right|_\rho - \left.\frac{\partial \ln \mathcal{C}}{\partial \ln \rho}\right|_T - \left.\frac{\partial \ln \mathcal{H}}{\partial \ln T}\right|_\rho + \left.\frac{\partial \ln \mathcal{H}}{\partial \ln \rho}\right|_T\right). \tag{7}$$

In deriving these expressions, we have made use of the perfect-gas law, as well as the equilibrium condition, $\mathcal{H} = \mathcal{C}$. Equations 1–5 must be valid in each medium.

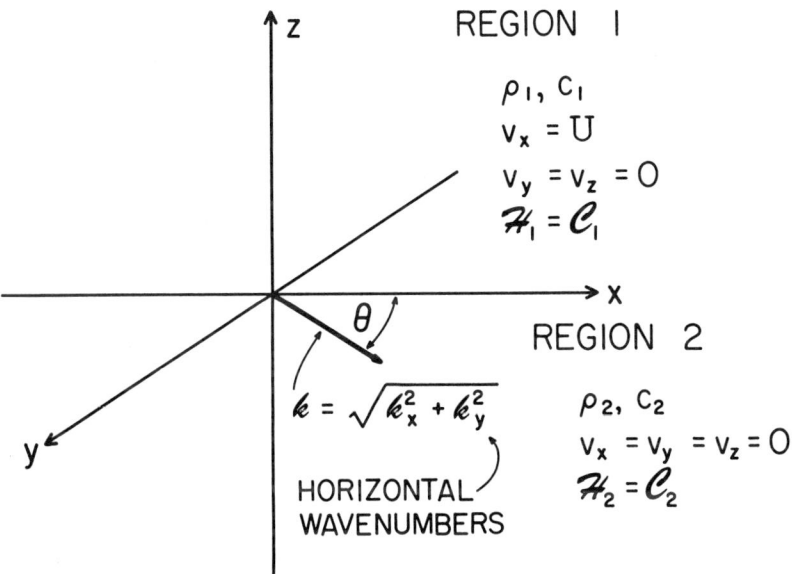

FIGURE 1. An illustration of the idealized Kelvin-Helmholtz problem.

We then Fourier analyze the disturbances, letting any perturbed quantity $\xi' = \xi'(z) \cdot \exp[i(k_x x + k_y y + nt)]$. By elimination, we can arrive at a single equation for any of the perturbations:

$$\left[D^2 - k^2 + \frac{\phi^2(\phi - i\mathcal{F}_p)}{c^2(\phi - i\mathcal{F}_\rho)}\right]\xi'(z) = 0, \tag{8}$$

where $D^2 = d^2/dz^2$ and $\phi = n + k_x U$. The eigenfunctions may be written as $\xi'(z) = A \cdot e^{qz}$, where A is constant, if q satisfies the expression,

$$q^2 = k^2\left[1 - \frac{\phi^2(\phi - i\mathcal{F}_p)}{k^2 c^2(\phi - i\mathcal{F}_\rho)}\right]. \tag{9}$$

Unless q^2 is real and <0, the solutions of this equation are of the forms, $q = \pm k(d + if)$, where $d > 0$. (If q^2 is real and $> 0, f = 0$.) We define a dimensionless growth rate and other dimensionless parameters as follows: $x \equiv n/(kc_1)$, $\alpha \equiv \mathcal{F}_p/(kc_1)$, $\beta \equiv \mathcal{F}_p/(kc_1)$, and $b \equiv (c_1/c_2)^2$. Thus, $\phi_1 = x + m$ and $\phi_2 = x$. Because the disturbances are driven by the energy (and momentum) flow near the interface, we require that they vanish as $z \rightarrow \pm\infty$. Hence, in regions 1 and 2, respectively, $q_1 = -k(d_1 + if_1)$ and $q_2 = +k(d_2 + if_2)$, unless q^2 is real and <0. If the q's are imaginary, then Sommerfeld (radiation) boundary conditions[7] must be applied, which allow only outgoing waves as $z \rightarrow \pm\infty$.

A perturbed displacement normal to the interface z' is related to w' by the expression $\mathcal{D}z' = w'$. Hence, from equation 4,

$$\rho\mathcal{D}[\mathcal{D}z'] \equiv \rho\mathcal{D}^2 z' = \frac{-\partial p'}{\partial z} = -qp', \tag{10}$$

which is a result valid in both media.

At the interface between the media, both p' and z' must be continuous. Therefore,

$$\frac{\rho_1 \phi_1^2}{q_1} = \frac{\rho_2 \phi_2^2}{q_2} \tag{11}$$

or

$$\frac{a(x+m)^2}{\left\{1 - (x+m)^2 \left[\frac{(x+m) - i\alpha_1}{(x+m) - i\beta_1}\right]\right\}^{1/2}} = \frac{-x^2}{\left\{1 - bx^2 \frac{(x - i\alpha_2)}{(x - i\beta_2)}\right\}^{1/2}}, \tag{12}$$

where $a = \rho_1/\rho_2$ and where it is understood that the real parts of the square roots of the complex terms are >0. Equation 12 is the fundamental dispersion relation (or characteristic equation) for anisotropic, K-H–like instabilities. An equation of this same form is valid in the transverse ($k = k_1$) hydromagnetic case if the unperturbed magnetic field is parallel to the y-axis.[9]

PARTICULAR SOLUTIONS AND PRELIMINARY RESULTS

When squared and cleared of fractions, the expanded version of equation 12 becomes an eighth-order polynomial in x with complex coefficients that cannot be solved algebraically. In the adiabatic case ($\alpha_{1,2} = \beta_{1,2} = 0$), it reduces to a sixth-order equation with real coefficients. However, in order for an equilibrium configuration to be realized, the unperturbed pressure must be continuous across the interface. Hence, $\rho_1 c_1^2 = \rho_2 c_2^2$ or $ab = 1$. After imposing this constraint, Hunter[9] has shown that the maximum unstable growth occurs in the adiabatic limit when $\rho_1 = \rho_2$, or $a = 1$, just as is true for incompressible fluids.[9] Therefore, in this preliminary study, we confine our attention to cases in which both fluids are identical: $a = b = 1$, $\alpha_1 = \alpha_2$, and $\beta_1 = \beta_2$. The adiabatic version of this problem was first considered by Miles[8] and Gerwin.[7]

The expanded, adiabatic dispersion relation may be written in the form

$$\left(x + \frac{m}{2}\right)[(x^2 + mx - 1)^2 - (m^2 + 1)] = 0. \tag{13}$$

The roots of this equation are

$$x_{1,2} = -\frac{m}{2} \pm i\left\{\sqrt{m^2 + 1} - 1 - \frac{m^2}{4}\right\}^{1/2}, \quad (14a)$$

$$x_{3,4} = -\frac{m}{2} \pm \left\{\sqrt{m^2 + 1} + 1 + \frac{m^2}{4}\right\}^{1/2}, \quad (14b)$$

and

$$x_5 = -\frac{m}{2}. \quad (14c)$$

Complex solutions $x_{1,2}$ represent the K-H modes; the imaginary part of the unstable K-H root bears the $(-)$ sign. Real solution x_5 has been designated a new wave by Gerwin.[7] Wave solutions $x_{3,4}$ do not satisfy the radiation boundary conditions.[7] The K-H roots are valid solutions for all m, whereas x_5 is valid for $m > 2$. K-H instabilities grow most rapidly when $m = \sqrt{3}$, and they exist for $m < \sqrt{8}$. We refer to the latter regime as the Mach Envelope. For $m > \sqrt{8}$, the K-H modes are purely oscillatory.

Thermal parameters, α and β, are proportional to t_s/t_c. Hence, these quantities are large for short cooling times, as is often the case in the ISM. As an example, consider a perturbation of wavelength $\lambda = 10^{19}$ cm (~ 3 parsecs) in an HI region having $T = 10^2$ K and a mean molecular weight of 1.4. The sound speed is $\sim 10^4 T^{1/2} = 10^5$ cm s^{-1} and therefore $t_s = \lambda/c \sim 10^{14}$ s. Using Spitzer's[11] cooling rates and assuming a low rate of ionization ($\leq 10^{-3}$), $t_c \sim 10^{13}/n$ s, where n is the number density. Therefore, $t_s/t_c \sim 10n$, which can easily be of order 10^2–10^3 in moderately dense HI regions. This ratio may assume even larger values in molecular clouds. In contrast, moderate to small values of α and β may arise in very tenuous HI regions or for disturbances of much shorter wavelength (or both).

When $|\alpha|$ and $|\beta| \rightarrow \infty$, there are two classes of asymptotic solutions. The first is designated Class 1: $\alpha/\beta > 0$ and both α and $\beta \rightarrow \infty$. Unperturbed models of this class would be thermally stable according to the usual, static criteria.[5,6] Let $\tilde{x} = \sqrt{\alpha/\beta} \cdot x$ and $\tilde{m} = \sqrt{\alpha/\beta} \cdot m$. In this limit, the dispersion relation becomes

$$\left(\tilde{x} + \frac{\tilde{m}}{2}\right)[(\tilde{x}^2 + \tilde{m}\tilde{x} - 1)^2 - (\tilde{m}^2 + 1)] = 0, \quad (15a)$$

which is exactly the same form as in the adiabatic case. Consequently, the acceptable ("adiabatic") solutions are

$$x_{1,2} = -\frac{m}{2} \pm \sqrt{\frac{\beta}{\alpha}}\left\{\sqrt{\frac{\alpha m^2}{\beta} + 1} - 1 - \frac{(\alpha m^2)}{4\beta}\right\}^{1/2}, \quad (15b)$$

and

$$x_5 = -\frac{m}{2}. \quad (15c)$$

The new Mach Envelope is defined by $0 < m < \sqrt{8\beta/\alpha}$.

The second class is designated Class 2: $\alpha/\beta < 0$ and both $|\alpha|$ and $|\beta| \rightarrow \infty$. Unperturbed models of this class would be thermally unstable according to the usual criteria and, therefore, it is not clear what significance should be attached to the instabilities discussed below. Let $\tilde{x} = \sqrt{|\alpha|/|\beta|} \cdot x$ and $\tilde{m} = \sqrt{|\alpha|/|\beta|} \cdot m$. In these variables, the dispersion relation reads

$$\left(\tilde{x} + \frac{\tilde{m}}{2}\right)[(\tilde{x}^2 + \tilde{m}\tilde{x} + 1)^2 + (\tilde{m}^2 - 1)] = 0. \quad (16)$$

The K-H solutions are

$$x_{1,2} = -\frac{m}{2} \pm i\sqrt{\left|\frac{\beta}{\alpha}\right|}\left\{1 - \sqrt{1 - \left|\frac{\alpha}{\beta}\right|m^2} - \left|\frac{\alpha}{4\beta}\right|m^2\right\}^{1/2}. \quad (17a)$$

In this limit, the new wave solution ($x_5 = -m/2$) does not satisfy the fundamental dispersion relation (equation 12).

The other complex roots are valid solutions as well:

$$x_{3,4} = -\frac{m}{2} \pm i\sqrt{\left|\frac{\beta}{\alpha}\right|}\left\{1 + \sqrt{1 - \left|\frac{\alpha}{\beta}\right|m^2} - \left|\frac{\alpha}{4\beta}\right|m^2\right\}^{1/2}. \quad (17b)$$

For this class of models, the Mach Envelope does not exist and complex instabilities occur at all Mach numbers.

In the preceding arguments, two thermal modes have been suppressed by the asymptotic nature of the analyses. If $\alpha/\beta > 0$, these modes are uninteresting because neither satisfies the dispersion relation. However, when $\alpha/\beta < 0$, one of the thermal modes is a valid solution for all m. Moreover, this mode is always unstable and it has a growth rate of the order of $|\alpha|$ and $|\beta|$. We interpret this result to mean that the undisturbed models of this class are not realistic when the cooling times are very short. Consequently, such models will not be considered further in the present communication.

When α and β are very small, but nonzero, we develop first-order corrections to any of the valid adiabatic solutions (equations 14a and 14c). We quote only the growth rate, x', of the new wave solution:

$$x' \approx \frac{im^2(\alpha - \beta)}{2(8 - m^2)}. \quad (18)$$

Consequently, when the cooling times are relatively long (which would be true for sufficiently tenuous HI regions), the new wave would be unstable when $\alpha < \beta$ and $m < \sqrt{8}$, as well as for when $\alpha > \beta$ and $m > \sqrt{8}$. As $m \to \infty$, $x' \to -i(\alpha - \beta)/2$. According to our first-order expansion, x' for the new wave solution has a resonance at $m = \sqrt{8}$, which is the outer boundary of the Mach Envelope for K-H instabilities. The appearance of the "resonance" does not mean that the absolute value of $x' \to \infty$ as $m \to \sqrt{8}$. Instead, x' becomes of the order of x near resonance; that is, when α and β are small, the growth rate of the new wave has a local maximum and minimum near $m = \sqrt{8}$.

Similarly, when both α and β are very large and positive, corrections can be developed to the valid "adiabatic" roots (equations 15b and 15c). In this limit, the correction to the new wave solution is

$$x' \approx \frac{im^4(\alpha - \beta)}{8\beta^2\left(8 - \dfrac{\alpha m^2}{\beta}\right)}.\quad(19)$$

A resonance appears at $m = \sqrt{8\beta/\alpha}$, which is the limit of the Mach Envelope when both α and $\beta \to \infty$. This solution is unstable either when $\alpha < \beta$ and $m < \sqrt{8\beta/\alpha}$, or when $\alpha > \beta$ and $m > \sqrt{8\beta/\alpha}$. As $m \to \infty$, $x' \to -im^2(\alpha - \beta)/(8\alpha\beta)$. Hence, for large m, we anticipate that the new wave should become progressively more unstable as m increases. This prediction has been verified by numerical solutions of the full expanded,

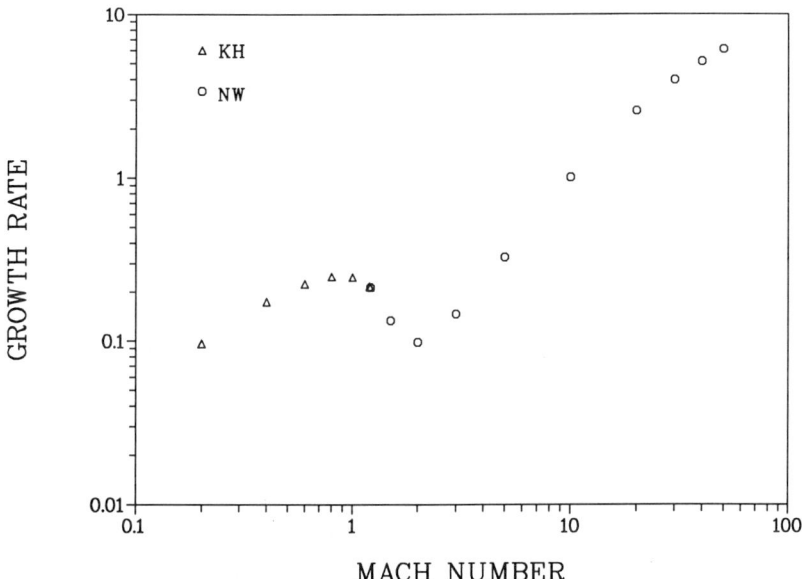

FIGURE 2. The dimensionless growth rates of the unstable Kelvin-Helmholtz (KH) and new wave (NW) modes as a function of effective Mach number when $\alpha = 30.00$ and $\beta = 7.143$.

characteristic equation. A detailed study is currently in progress concerning the behavior of the various instabilities as a function of the seven parameters: $a, b, \alpha_{1,2}, \beta_{1,2}$, and m. We have confirmed that the growth rate of the new wave changes sign near resonance and that the numerical solutions agree with our asymptotic formulae (equations 18 and 19) when $|\alpha|$ and $|\beta|$ are small and large. However, the implications of resonance phenomena in more realistic cases have yet to be explored. Our preliminary results indicate that there is a regime in (α, β, m) space in which each of the valid solutions would be expected to dominate.

In FIGURE 2, we illustrate the growth rates of the unstable modes as a function of m when $\alpha = 30.00$ and $\beta = 7.143$. These parameters would roughly characterize optically thin cooling in a molecular cloud[12] when $\lambda = 10^{19}$ cm, when the total molecular number density $= 10^2$ cm^{-3}, and when the heating occurs at a constant rate per gram ($\mathcal{H} = $ constant $\times \rho$). Within the Mach Envelope, a modified K-H instability occurs, whereas the new wave is unstable for all larger m values. At large m, the growth rate agrees closely with our asymptotic formula and a vestige of the resonance behavior is discernible in the range $1.2 \lesssim m \lesssim 2.0$. A dramatic growth of x' near resonance is shown in FIGURE 3. In this near asymptotic case, for which $\alpha = 3 \times 10^6$ and $\beta = 7.143 \times 10^5$, the very small growth rate of the new wave is amplified by $\sim 10^3$ at resonance and the discontinuous change from stable to unstable behavior is apparent at the boundary of the Mach Envelope.

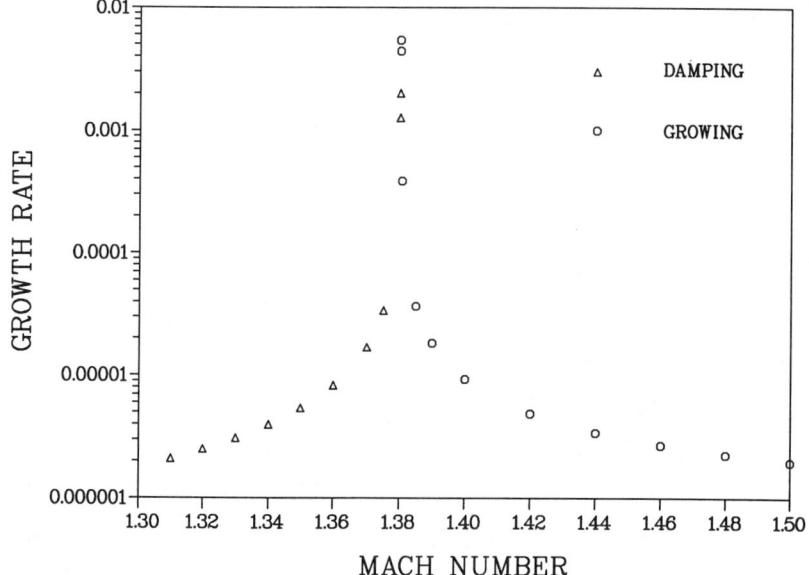

FIGURE 3. The absolute values of the dimensionless growth rates of the new wave near resonance as a function of effective Mach number when $\alpha = 3.0 \times 10^6$ and $\beta = 7.143 \times 10^5$. The damping rates are intrinsically negative.

One additional formal point is worth mentioning. If the velocity profile were not discontinuous (e.g., if it were linear or of the hyperbolic tangent form), then a more natural choice of reference frame would be one in which the fluid at large z approaches velocity $+U/2$ in the x direction, and in which the fluid situated at large negative z approaches an x velocity of $-U/2$. In such a symmetric frame, the dimensionless growth rates of the modes would equal our calculated rates plus $m/2$. Thus, the oscillatory frequencies of particular modes, such as the new wave, would vanish, and their growth rates would become pure exponentials. Notwithstanding, some modes must always oscillate because it is not possible to transform to a coordinate system in which the real parts of all roots vanish.

CONCLUSIONS

From the present work, we conclude the following:

(1) Mixed thermal, Kelvin-Helmholtz (K-H–like) instabilities can arise in the ISM at all Mach numbers.
(2) These instabilities, which are driven by both dynamical and thermal forces, would occur naturally in the ISM, even when each fluid would be thermally stable if treated separately (α and $\beta > 0$).
(3) In order to predict the behavior of such instabilities, it is crucial that the local heating and cooling mechanisms in the ISM be known in detail.

When the present, highly idealized theory is considered along with our previous work on Raleigh-Taylor–like instabilities in colliding interstellar gas flows,[3] we speculate that hybrid Raleigh-Taylor, K-H–like instabilities are responsible for much of the disorder that is observed in the ISM.

REFERENCES

1. SPITZER, L. 1978. Physical Processes in the Interstellar Medium, p. 144. Wiley. New York.
2. MYERS, P. C. 1985. Molecular cloud cores. *In* Protostars and Planets II. D. C. Black & M. S. Matthews, Eds.: 81–103. Univ. of Arizona Press. Tucson, Arizona.
3. HUNTER, J. H., JR., M. T. SANDFORD II, R. W. WHITAKER & R. I. KLEIN. 1986. Star formation in colliding gas flows. Astrophys. J. **305**: 309–332.
4. HUNTER, J. H., JR. & R. C. FLECK. 1982. Star formation: The influence of velocity fields and turbulence. Astrophys. J. **256**: 505–513.
5. FIELD, G. B. 1965. Thermal instability. Astrophys. J. **142**: 531–562.
6. HUNTER, J. H., JR. & S. SOFIA. 1971. The dynamics and thermal stability of planetary nebulae. Mon. Not. R. Astron. Soc. **154**: 393–413.
7. GERWIN, R. A. 1968. Stability of the interface between two fluids in relative motion. Rev. Mod. Phys. **40**(3): 652–655.
8. MILES, J. W. 1959. On the disturbed motion of a plane vortex sheet. J. Fluid Mech. **6**: 538–552.
9. HUNTER, J. H., JR. 1985. Kingfish striations and the Kelvin-Helmholtz instability: part I. Los Alamos Report (LA-10566-MS).
10. BLUMEN, W., P. G. DRAZIN & D. F. BILLINGS. 1975. Shear layer instability of an inviscid compressible fluid. Part 2. J. Fluid Mech. **71**: 305–316.
11. SPITZER, L. 1978. Physical Processes in the Interstellar Medium, p. 143. Wiley. New York.
12. HOLLENBACH, D. J. & C. F. MCKEE. 1979. Molecule formation in fast shocks. Astrophys. J. Suppl. Ser. **41**: 577–581.

Index of Contributors

Bartlett, J. H., 78–82
Bensimon, D., 110–116
Buchler, J-R., ix–xi, 37–54

Coffey, S. L., 22–36
Contopoulos, G., 1–15

Deprit, A., 22–36

Eichhorn, H., ix–xi

Fry, J. N., 66–77

Grebogi, C., 117–126

Hunter, J. H., Jr., 144–153

Kadanoff, L. P., 110–116
Koszykowski, M. L., 127–143

Meiss, J. D., 83–96
Miller, B., 22–36

Noid, D. W., 127–143
Nusse, H. E., 117–126

Ott, E., 117–126

Pfeffer, G. A., 127–143

Schmidt, G., 97–109
Schwarzschild, M., 16–21
Smith, L. A., 61–65
Spiegel, E. A., 55–60, 61–65

Whitaker, R. W., 144–153
Williams, C. A., 22–36
Wolf, A., 55–60

Yorke, J. A., 117–126